Praise for *A N*

"With his special gift for revealin
historical characters, Kenneth Davi ered, haunt-
ing narrative. Peeling back the vene self-serving nineteenth-
century patriotism, Davis evokes the raw and violent spirit not just
of an 'expanding nation,' but [also] of an emerging and aggressive
empire." —Ray Raphael, author of *Founders*

Praise for *America's Hidden History*

"American history in the vibrant narrative tradition of David
McCullough." —Ron Powers, author of *Flags of Our Fathers*

"Paul Revere didn't shout out 'The British are coming!' as he
galloped through the night on his famous ride. What did he actually say? Read Kenneth Davis's *America's Hidden History* and you'll
find out. You'll learn lots more, too, in this engaging book."

—Joy Hakim, author of *A History of US*

Praise for *Don't Know Much About® History*

"*Don't Know Much About® History* . . . puts the zest back in history."

—*Washington Post Book World*

"Fun, engrossing, and significant. . . . History in Davis's hands is
loud, coarse, painful, funny, irreverent—and memorable."

—*San Francisco Chronicle*

Also by Kenneth C. Davis

America's Hidden History: Untold Tales of the First Pilgrims, Fighting Women, and Forgotten Founders Who Shaped a Nation

Don't Know Much About History

Don't Know Much About Geography

Don't Know Much About the Civil War

Don't Know Much About the Bible

Don't Know Much About the Universe

Don't Know Much About Mythology

Two-Bit Culture: The Paperbacking of America

A Nation Rising

Untold Tales from America's Hidden History

Kenneth C. Davis

HARPER

NEW YORK . LONDON . TORONTO . SYDNEY

HARPER

A hardcover edition of this book was published in 2010
by HarperCollins Publishers.

For information address HarperCollins Publishers, 10 East 53rd Street,
New York, NY 10022.

HarperCollins books may be purchased for educational, business, or sales promotional use. For information please write: Special Markets Department, HarperCollins Publishers, 10 East 53rd Street, New York, NY 10022.

FIRST HARPER PAPERBACK PUBLISHED 2011.

Designed by Suet Yee Chong

The Library of Congress has catalogued the hardcover edition as follows:
Davis, Kenneth C.
A nation rising: untold tales of flawed founders, fallen heroes, and forgotten fighters from America's hidden history / by Kenneth C. Davis. — 1st ed.
p. cm.
Includes bibliographical references and index.
ISBN: 978-0-06-111820-3
1. United States—History—1783–1815—Biography—Anecdotes. 2. United States—History—1815–1861—Biography—Anecdotes. I. Title.
E339.D38 2010
973.09'9—dc22 2009049641

ISBN 978-0-06-111821-0 (pbk.)

11 12 13 14 15 OV/RRD 10 9 8 7 6 5 4 3 2 1

For my children, Jenny and Colin Davis

A rising nation, spread over a wide and fruitful land, traversing all the seas with the rich productions of their industry . . . advancing rapidly to destinies beyond the reach of mortal eye.

—Thomas Jefferson,
First Inaugural Address (March 1801)

These were our founders: imperfect men in a less than perfect nation, grasping at opportunities. That they did good for their country is understood, and worth our celebration; that they were also jealous, resentful, self-protective, and covetous politicians should be no less a part of their collective biography. What separates history from myth is that history takes in the whole picture, whereas myth averts our eyes from the truth when it turns men into heroes and gods.

—Nancy Isenberg,
Fallen Founder

CONTENTS

Introduction

"The Dream of Our Founders"

It would be difficult—no, it would be impossible—to have witnessed the events surrounding Election Day 2008 and Inauguration Day 2009, either as a historian, as an interested observer, or simply as an American, and not to have been profoundly struck by their place in our history. The stunning election of Barack Obama has rightfully been judged a transforming moment which historians may someday rank alongside the elections of Andrew Jackson, Abraham Lincoln, Franklin D. Roosevelt, John F. Kennedy, and Ronald Reagan—fulcrum moments in which the course of American history took a sharp, sudden, and decisive turn. And it would be equally difficult, if not impossible, to continue to write about the shaping of a nation without taking into account this extraordinary milestone in American history.

Obama's election to the nation's highest office marked a profound reversal of many long-held assumptions about geography, gender, parties, politics, and race relations—indeed, the American character itself—that have been entwined in this nation's fabric since the arrival of Europeans in North America more than 400 years ago.

On the night of his election in 2008, Obama offered a victory speech touching on this upheaval of the American political landscape and its place in the drama of American history. Speaking in Chicago's Grant Park to a throng of deliriously joyful and tearful celebrants, the forty-seven-year-old president-elect opened by saying:

"If there is anyone out there who still doubts that America is a place where all things are possible; who still wonders if the dream of our founders is alive in our time; who still questions the power of our democracy, tonight is your answer."

"THE DREAM OF OUR FOUNDERS"

GRANTING THE FUNDAMENTAL notion that America is a place of great opportunity, it is still nearly impossible to contemplate Obama's phrase, "the dream of our founders," without pondering its extraordinary and obvious corollary: many of those dreaming founders would have been perfectly at home owning Barack Obama, his wife Michelle, and their two little girls and perhaps selling all or some of them—either for profit or to pay off debts. That august group would include, of course, George Washington and Thomas Jefferson, along with many of the Founders who signed the Declaration of Independence and the Framers who wrote the United

States Constitution. At the time of its creation in Philadelphia, the Constitution stated that had Barack Obama then been a "person held in service," he would have been counted as "three-fifths of a man" for the purposes of allotting seats in Congress.

So the Founders' dream of "Life, Liberty, and the Pursuit of Happiness" is real and possible. But it has always existed uneasily alongside the insidious influence that the legacy of slavery and race has exerted on America's past—the stunning gap between America's ideals and its realities. These two competing visions of *e pluribus unum* lie at the heart of this country's great contradiction, and they frame the moment of Obama's election.

The momentous upheaval brought about by Obama's victory did not wipe the slate clean. For all the distance that America has traveled as a nation since 1776, the country still needs to reconcile the glorious dream with the dark nightmare that haunts America's past. And the sharp contradiction that pits the history of slavery and race against "the dream of our founders" is nowhere more clearly and devastatingly laid bare than in the period covered in this book, the crucial first fifty years of the nineteenth century.

It was a dynamic and dramatic half century during which the United States changed with stunning speed from a tiny, newborn nation, desperately struggling for survival on the Atlantic seaboard, to a near-empire spanning the continent, "from sea to shining sea." In 1800, according to the census, the United States population stood at 5,308,483, of which 893,602 were slaves. With the Louisiana Purchase in 1803, the nation doubled in size geographically. By 1826 and the "Jubilee" celebration of the fiftieth anniversary of the

Declaration of Independence, the population, swollen by waves of immigrants, had reached more than 12 million. And by 1850, with an even greater influx of immigrants, it had nearly doubled again, to 23,191,876, with 3,204,313 of them slaves.

It was also a half century of tremendous technological innovation: canals, steamships, railroads, and the telegraph had made for revolutions in travel and communications. The sewing machine and photography changed everyday life. When the nineteenth century began, written messages and armies moved as they had for thousands of years. By the middle of that century, information was sent over telegraph lines in minutes. Steam revolutionized the movement of people, armies, and goods.

America began to be transformed in other less material, less utilitarian ways as well. This was the era that saw an emergence of "homegrown" American arts and letters. Breaking free from the strictures of British and European convention, a new generation of American writers created a distinctive American voice: Emerson, Hawthorne, Melville, Cooper, Whittier, Longfellow, Washington Irving, Poe, and that most original of voices, Walt Whitman—who even said it outright: "I hear America singing." All injected vitality into American letters. And many would cross the line between art and politics—actively writing, often in harshly critical terms, about the American scene they observed.

It was a time, in other words, when the "dream of our founders" was realized in ways that few men of the Revolutionary generation could have possibly imagined. And it was an era in which an inexorable succession of events ultimately led to the great, tragic conflagration that followed—the American Civil War.

But for all too many Americans, the events of those fifty-odd years—counting roughly from Thomas Jefferson's controversial election in 1800 to California's statehood in 1850—have fallen into a black hole called "American history." Even though some of us probably have dim recollections of such household names and phrases as the Non-Intercourse Act (a perennial high school favorite), the War of 1812 (What was *that* about?), Manifest Destiny, the Missouri Compromise, "Tippecanoe and Tyler too," the Mexican War (What was *that* about?), and the California gold rush, the details are frequently fuzzy. Far too many people skip from 1776 to the Civil War without a clear sense of what happened during a good part of Lincoln's famous "four score and seven years."

And that's too bad, because this era stands as one of the most extraordinary and tumultuous periods in America's relatively brief history. When we move beyond the nineteenth-century household names and textbook buzzwords such as nullification, states' rights, or manumission, the far-reaching human impact of these crucial decades becomes even more compelling. So I return to America's "hidden history"—the obscure, forgotten or deliberately erased events and people that had a profound impact on the shaping of a nation.

As I attempted to do in *America's Hidden History*, I have set out to distill a segment of American history through six separate narratives, each of which focuses on a largely overlooked or untold incident that helped decide or at least influence the nation's course. Each of these stories provides a portal into the times in which it took place, the issues people faced, the sacrifices that were made, and the blood that was spilled. Again, these pictures of an overlooked America are not always pretty, and often are at odds with the comfortable notions so many people still cling to about America's pioneering past.

Again, I have chosen to focus on the human side of the story. These events are peopled by both the famous and the forgotten, by many once familiar names. The events highlight the now obscure and the notorious. They are tales of conquest, conspiracy, corruption—and courage. Each played some part in shaping the nation's geography and destiny.

Does the name William Weatherford ring a bell? No? How about Francis Dade? Then there are Madison Washington and Jessie Frémont.

Swept into the dustbin of American history, these names belong to people who occupied, at least for a time, the day's front pages. They were involved in crises and controversies that affected the direction taken by the new nation. But today they are mostly forgotten—or, like Aaron Burr, reduced to a single note in an encyclopedia: "Jefferson's first vice president. Killed Alexander Hamilton in a duel."

Each of these stories also explores a vastly more complex path to nationhood than the tidily packaged national myth of rugged individuals setting out for new territory. It is a path of confusing and competing alliances and ambitions—the ambitions of individuals and the ambitions of a nation poised to take its place on a grand world stage.

The story of the "rising nation" described in this book is bracketed by two chapters describing a pair of high-profile trials, both of which featured some of the most prominent men of the nineteenth century. In an era that lacked twenty-four-hour cable television and cameras in courtrooms, both of these "trials of the century" captivated the nation. In the first, Aaron Burr was tried, not for his role in the death of Alexander Hamilton, but for treason. His trial demonstrated that, even in 1807, presidents could be vindictive and were not hesitant about destroying their opponents in the press and using the power of the government and the courts for political ends.

More than 200 years after Burr's trial, the danger of presidential vendettas is still real. Richard Nixon compiled an "enemies list" and used the power of the FBI and the CIA to break people who were perceived as threats, such as Daniel Ellsberg, the defense analyst who leaked the "Pentagon Papers." More recently, we have witnessed the power of the White House in attacking personal reputations through the use of media "leaks," as in the case of Ambassador Joseph Wilson, a critic of the Iraq War, and his wife, Valerie Plame, a CIA operative who was "outed" by members of the George W. Bush administration.

My interest in Burr is similar to my fascination with the character and career of Benedict Arnold, whose complex rise and fall I explored in *America's Hidden History*. Arnold and Burr's shared reputation as the "bad boys" of American history is intriguing. Burr's character became even more alluring to me when I learned that, as a nineteen-year-old soldier, Burr had gone off to war with Benedict Arnold, tramping through the Maine woods in 1775. Oh, to be an observer of their exchanges as Arnold led a dreadful forced march during which starvation, disease, and desertion had whittled a force of 1,000 men down to fewer than 700.

In reality, the so-called villains of history often make the most fascinating subjects. Burr's trial for treason also opened the way to many of the larger themes touched upon in this book's other chapters—the voracious American appetite for land; the conflict with Spain, which still held enormous tracts of what would eventually be American territory; and the powerful forces of greed and ambition that drove so many men at this time.

And just as there was a link between Benedict Arnold and Aaron Burr, I found a connection between Burr and another important character in this book, Andrew Jackson. The future president was cast in a relatively minor part in Burr's trial, but Jackson reappears to play a central role in several of these chapters. And that is why exploring this "hidden history" is so compelling to me.

Over the years, Jackson has received favorable reviews as a president. Frequently ranked among the most influential commanders in chief, he was recently lionized in Jon Meacham's Pulitzer Prize-winning bestseller about his presidency, the aptly titled *American Lion*.

But not everyone admires Andrew Jackson. There is more to the story. In his recent book *Waking Giant*, David S. Reynolds described the seventh president as "a potent killing machine." And that is where the "hidden history" comes in. There are many people, mostly descendants of Native Americans, who won't accept a $20 bill, because they do not want to use money bearing Andrew Jackson's image. In fact, because of Jackson's draconian policies toward Native Americans and his devotion to the cause of slaveholding, there is a small movement afoot to replace Jackson on the $20 bill.

The second trial, which closes the book, recounts the very public life of John Charles Frémont—soldier, explorer, and relentless self-promoter, once a larger-than-life American hero who was lauded across the land for his pivotal role in opening the American West to pioneers. He and his extraordinary wife, Jessie Benton Frémont, were once the nation's most famous couple, a nineteenth-century version of "Jack and Jackie." Set against the drama of Manifest Destiny and California's entry into the Union, Frémont's trial capped a period of political infighting and intrigue during which the man famous as "the great pathfinder" faced a court-martial in what one historian called "a case study in the dynamics of reputation." A few years later, in 1856, Frémont emerged as the Republican Party's first presidential candidate on an antislavery platform, as America edged precipitously closer to the Civil War.

Between these two very famous Americans are stories of some less familiar names and events. A fitting example is William Weatherford and the massacre at Fort Mims. This incident, which ranks as the worst massacre in American frontier history, was as shocking and devastating an attack in its own time as the terrorist strikes of 9/11 were in ours. And this episode, and the little-known war it sparked, began to form Andrew Jackson's heroic national reputation.

The story of Fort Mims speaks volumes about the tangled politics of the time, the overarching American crusade for more territory to add to the nation's holdings, and the complex web of relationships at the heart of that strife—Americans, British, French, Spanish, Indians, and America's blacks, both slave and free, all in deadly contention for land, freedom, and survival.

That pair of linked struggles over land and freedom—in essence, the true "dream of our founders"—resonates through the stories recounted here. Take the nearly fifty-year war in Florida between the American government and the Seminoles, allied with runaway slaves and free blacks. A piece of that conflict's history is detailed in "Dade's Promise," a reference to the ill-fated army commander Francis Dade, for whom Dade County is named. This attack by Seminole Indians on American troops—another nineteenth-century counterpart of 9/11—led to the longest war in American history until Vietnam. Of course, most Americans today are familiar with the "Seminoles" only as the nickname of the Florida State University sports teams.

The very human face of the struggle for freedom is also on display in the tale of Madison Washington. An escaped slave who returned to the South in a vain attempt to free his wife from bondage, Madison Washington was recaptured and resold before leading a mutiny aboard a slave ship. This little-noted episode provides a glimpse into the centuries of insurrections and rebellions that occurred throughout America's slaveholding history—an account much at odds with the traditional mythic view of complacent slaves accepting their bondage. And that too is an essential piece of America's "hidden history."

Another example of the mythic past continuing to haunt the present is on display in the account of the "Bible Riots," a paroxysm of deadly sectarian fighting in the 1840s that shattered the peace of Philadelphia, the City of Brotherly Love. This tableau of American intolerance and religious violence should completely shatter the comfortable notion of a so-called Christian nation. Many Americans shake their heads at the scenes of pitched street fighting in such places as Baghdad or Rwanda, wondering how countrymen from rival sects or tribes can brutalize one another. But it happened here too. The virulent anti-immigrant fever that swept through America during much of the nation's early history stands in sharp contrast to the "melting pot"—a myth that has been such a large part of the American dream. It is a stark reminder that the nation's immigration policies have always been and remain one of the hottest of hot-button issues.

Each of these "untold tales" reflects on the basic idea of how America truly came to be the nation it was by the mid-nineteenth century—and in many ways continues to be. The issues raised by these stories—ambition, power, territorial expansion, slavery, intolerance, the rights of the accused, the use of the press, disdain for the immigrant—all continue to reverberate in our headlines. So do the uses of fear and propaganda, which have been components of the American story from the country's earliest days.

In uncovering these hidden stories, I have found some uncanny parallels to contemporary American politics. For instance, there is nothing new about "birthers," those people convinced that President Obama was born in Kenya. In 1856, the presidential candidate John Frémont had to face the widespread rumor that he was French-Canadian by birth. And the "tea party" movement spawned in 2009 in homage to the Boston Tea Party is nothing new. In the 1840s, "tea baggers" burned down a Catholic convent.

These stories present a different truth, one that is often hidden from our view. Nearly fifty years ago, President John F. Kennedy said, "For the great enemy of truth is very often not the lie—deliberate, contrived and dishonest—but the myth—persistent, persuasive and unrealistic. Too often we hold to the clichés of our forebears. . . . We enjoy the comfort of opinion without the discomfort of thought."[1]

• • •

GETTING THE WHOLE "truth" and separating it from "myth" raises a question I often hear. Whether in a school, at a teachers' conference, or at a lecture in a public library, people ask why this history is "hidden." Why don't we teach our children, and ourselves, the truth?

While doing the research for this book, I stumbled across a perfect example of those well-meaning but misguided attempts to mask reality and varnish the truth. Traveling through Florida to investigate the region's Spanish past, I visited one of the state's historic sites: a re-created Spanish colonial settlement, complete with an Apalachee native village. It was staffed by articulate, enthusiastic museum workers, many in period costume, who were clearly passionate about bringing history to life. In its otherwise excellent educational center, I came across a wall-size timeline of Florida's history. In large, clear letters it noted that the Spanish had "banished" the French from Florida in 1565. Reading that, I could only ask myself in astonishment, "Since when is *banished* the same thing as *massacred*?"

The French Huguenots referred to in that timeline were all mercilessly put to death in a religious bloodbath. The destruction of Fort Caroline, near present-day Jacksonville, in September 1565 and the slaughter of French soldiers a few weeks later was a mass sectarian killing by the Spanish admiral who founded Saint Augustine. The museum's "banished" was the rough equivalent of a German display contending that Europe's Jews had been "deported."

But we can't tell *that* to the children, can we? Such things don't happen in America! So we allow them to pass through history class and let them walk through a museum "learning center," safe in the knowledge that they are spared the ugliness in our past. Of course, the problem is that they never learn the truth. Or if they do, they have a perfectly good reason for cynicism.

One simple answer is that we have been understandably ashamed of this picture, which is at odds with the airbrushed version of the past. "We gauge our prospects as a people by locating a past from which we can draw hope and pride," the historian Andrew Burstein once noted. "Heroes become necessary in such an enterprise. . . . There is another way to say this. Biography is never a faithful record. It is a construction, a clandestine effort to refashion memory, to create a new tradition, or sanction yet another myth about what is past."[2]

Or we are worried about our children, and we assume that they need to be protected from the violent nature of the truth. And so our schoolbooks omit the unpleasantness, wishfully thinking that we can craft a vision of the past filled with pride and patriotism. But truth is a harsh mistress.

For years, I have contended that our attempts to "clean up" the past was somewhat akin to removing the picture of the mad aunt or disgraced uncle from the family album. But sooner or later, the kids stumble across an old snapshot and then you have to explain the whole, awkward thing.

Admittedly, I have learned that telling the truth, the whole truth, and nothing but the truth about the American past is an irritant to some people. I have been accused of "tearing down" our heroes. But it seems to me that getting at the heart of where we have come from as a nation is the only proper way in which the country can ever hope, as Barack Obama once put it, "to narrow that gap between the promise of our ideals and the reality" of what this country has been through.

Of course, in the best and worst of times, it is comfortable to think that America has held fast to the "dream of our founders," a dream that has inspired millions of Americans to work and sacrifice and countless millions more to come to America. And it has breathed life into freedom movements around the world, all of them hoping to be part of that dream of "Life, Liberty, and the Pursuit of Happiness."

This was the "dream of our founders." But is the dream alive? Or worse, is it simply a pipe dream, destined to disappoint? That is a fair question to ask, as America's faith in the dream has been severely tested in the first decade of the twenty-first century by the toll of terrorism, a series of shocks to the financial system, and a collection of cataclysmic government failures.

From my own optimistic perspective, however, I will join one of the dreaming heroes of my youth, Robert F. Kennedy, in quoting George Bernard Shaw:

"You see things; and you say 'Why?' But I dream things that never were; and I say, 'Why not?'"

—Kenneth C. Davis
Dorset, Vermont
September 2009

I

Burr's Trial

1789 The first Congress convenes in New York City in March.

Washington begins an eight-day journey from Mount Vernon to New York City. On April 30, he is inaugurated at Federal Hall.

1790 Secretary of War Henry Knox signs a peace treaty with Creek Indians. With Spanish support, however, the Creeks resume attacks on American frontier settlements.

1792 George Washington is reelected president, with John Adams as vice president.

1793 Eli Whitney's cotton gin revolutionizes cotton production.

1796 Tennessee, a slave territory, is admitted to the Union as the sixteenth state.

John Adams wins the presidency; Thomas Jefferson finishes second and becomes vice president.

1799 George Washington dies at age sixty-seven on December 14.

1800 The federal government moves from Philadelphia to Washington, D.C., in June.

The balloting in the December presidential election produces a tie. The contest is sent to the House of Representatives.

1801 After thirty-six ballots, the House chooses Thomas Jefferson as third president of the United States on February 17; Aaron Burr becomes vice president.

1803 *Marbury v. Madison*: the Supreme Court under Chief Justice John Marshall rules an act of Congress null and void when it conflicts with the provisions of the Constitution, establishing the principal of "judicial review."

1803 The Louisiana Purchase. The United States roughly doubles in size. Under the terms of the purchase, the status of West Florida and Texas is left unclear.

1804 Burr and Hamilton duel on July 11. Alexander Hamilton dies the next day.

1805 Thomas Jefferson is inaugurated for a second term.

1807 Aaron Burr is indicted for treason on June 27.

As long as it is impossible for you to transact your business in person, if you repose no confidence in delegates, because there is a possibility of their abusing it, you can have no government; for the power of doing good is inseparable from that of doing some evil.

—John Marshall,
arguing for the Constitution, 1788

[A]s a public man, he is one of the worst sort—a friend to nothing but as it suits his interest and ambition. . . . 'Tis evident that he plans on putting himself at the head of what he calls the "popular party," as affording the best tools for an ambitious man to work with. . . . In a word, if we have an embryo Caesar in the United States, 'tis Burr.

—Alexander Hamilton

But in this instance, he wished to teach a lesson on political persecution, and to demonstrate that justice only existed when the lone individual could successfully confront the tyrannical hand of state power. He had already revealed his approach in a letter to [his daughter] Theodosia, asking her to compose an essay containing all the episodes in ancient history when "a man of virtue and independence and supposed to possess great talents," had become "the object of vindictive and unrelenting persecution."

—Nancy Isenberg,
Fallen Founder

Mississippi Territory

February 18, 1807

Two riders were approaching.

It was nearly midnight. Nicholas Perkins, an attorney and militia officer, was playing backgammon with a friend in the small town of Wakefield—to the north of Mobile—when he heard a horse and rider pass at a brisk trot. One set of hooves seemed to pass. But when a second rider followed and stopped, Perkins went to his door. The second rider asked him for directions to the home of a Major Hinson. The other rider reined in his horse and waited a short distance away.

Perkins briefly studied the silent rider, the smaller of the two horsemen. A beaver hat, pulled low, partly concealed his face. Perkins observed that the man's expensive riding boots did not match the rest of his outfit, which was rough-hewn. Knowing the difficulty of following the roads in a backwoods wilderness where the threat from Indians was very real, Perkins advised the two riders not to press on. Instead, he suggested that they spend the night at a nearby tavern.

Rejecting the advice, the two men galloped off. But the pair of riders and the circumstances struck Perkins as too unusual to dismiss, and their unwillingness to stop raised his suspicions. "Could they be robbers with 'a bad design' on Hinson and his property? he wondered. Then another thought crossed his mind. Might the mysterious rider be Aaron Burr, making his escape through this remote country?"[1]

This was not a random notion. The former vice president, Aaron Burr, was thought to be traveling in the territory. And Nicholas Perkins knew that a reward of $2,000 had been posted for his arrest, amid swirling rumors and newspapers' speculation that Burr was up to no good. For months, the air had been filled with whispers and reports that the former vice president was gathering an army to foment an uprising or a war. Then, in January, in a letter to Congress, President Jefferson had openly accused Burr of a treasonous conspiracy. Suspecting that the mysterious rider might indeed be Burr, Perkins set off to find the local sheriff. They rode to Major Hinson's house, and it was there that Perkins got a better look at the man who had raised alarm bells.

"He was shocked by his bizarre getup. He wore a slouching white hat with a broad brim, sported a long beard and checkered handkerchief around his neck, and a great, baggy coat tied with a belt. Hanging from the belt was a tin cup and a butcher's knife. The outfit did not fit the profile of the dapper Burr, known for his stylish dress and genteel manners. But something gave him away:

'His eyes,' attested Perkins. . . . He later testified in court that he had heard 'Mr. Burr's eyes mentioned as being remarkably keen, and this glance from him strengthened his suspicions.'"[2]

Perkins waited outside Hinson's house while the sheriff spoke with the men. He then watched as Burr and the other rider, whose name was Robert Ashley, emerged from the house and rode with the sheriff, who was helping the wanted Burr get away.

If the picture of a lawman aiding and abetting a known fugitive seems incongruous, there were good reasons that the people in this part of America—what was then deemed the American "frontier" in the future state of Alabama—were sympathetic to Aaron Burr and his rumored plans. According to one openly discussed rumor, Burr was planning an audacious plot to outfit an army, begin a war with Spain, and capture Florida. Another rumor had Burr plotting to march on Mexico and claim that territory. The wildest of these rumors had Burr taking over a large swath of American western territory, seceding from the Union, and setting up his own empire. The ambitious and unscrupulous former vice president, forced from power in Washington, planned to install himself as emperor, another Napoléon, with his beloved daughter Theodosia, as his heir, an empress in waiting.

There was no love for the Spanish in America, especially in these southern territories that bordered Florida. Many Americans already expressed long-simmering, apparently inborn hostility toward the Spanish. Born of the Reformation-era religious wars between England and Spain, and fueled by centuries of anti-Spanish, antipapist propaganda, this hot streak ran especially deep among the Protestant Scots-Irish who had moved into America's southern wilderness. Many of these Americans wanted to press farther south into the Florida territory held by Spain since the founding of Saint Augustine in 1565.

Spain still controlled neighboring Florida and its coastline, as well as vast stretches of the American South and West and Mexico. Complicating matters, the transfer of extensive portions of North America from France to America in 1803 with the Louisiana Purchase had left the boundaries of Spain's territory in the area very murky. At the direction of President Jefferson, troops were sent to the borders to pressure Spain to surrender or sell the territory, and a swelling tide of American settlers continued to flood into the area, soon overwhelming the relatively few Spanish residents.

Hoping to slow the flow of American frontiersmen encroaching on their borders, the Spanish began to form alliances with Indian tribes in the territory, still numerous and eager to protect their ancestral hunting lands.

Last, the Spanish, and especially their Jesuit priests, had a reputation among American southerners for encouraging slaves to run away.

Apart from these "geopolitical" concerns, there was also little love lost for Alexander Hamilton, whom Burr had killed in a duel nearly three years earlier. As George Washington's secretary of the treasury, Hamilton had created a plan for a new American economy and a banking and credit system that seemed—in the eyes of many Americans—to favor well-heeled easterners at the expense of the less genteel westerners and southerners. Hamilton's plan called for assumption of the states' debts by the federal government in order to create a stable credit market in which the nation could put its financial house in order and more efficiently regulate commerce. The settlers and farmers who were pressing the country's boundaries with an almost insatiable desire for territory saw Hamilton's "Report on Public Credit," issued in January 1790, as a threatening plan for a federally dominated economy. Southerners in particular viewed this plan, with its powerful central national bank under the control of the president, as a threat to their notions of an unfettered economy, with each state controlling is own destiny.

More controversially, Hamilton's plan would reward financial speculators who had purchased Revolutionary War–era bonds and promissory notes, given to soldiers when the nascent United States literally could not pay its troops, at pennies on the dollar. Desperate for cash, many farmers, often veterans of the Revolution who thought the bonds worthless, sold most of these notes at depressed prices. Under Hamilton's plan, the speculators—who had admittedly taken a risk—would reap a windfall, and the "ordinary people" who had held the original bonds would lose out. This was the opening of a split between the investor class and the working class that was a harbinger of a dichotomy that still remains a powerful force in American politics and finance: Main Street versus Wall Street.

In his biography of Hamilton, Ron Chernow encapsulated the anti-Hamilton view:

> *Compounding Hamilton's problems was that his report crystallized latent divisions between north and south. There was a popular conception (to Hamilton, a gross misconception) that the original holders of government paper were disproportionately from the south and that*

the current owners who had "swindled" them were from the north. Hamilton denied that any such regional transfer took place. . . . Still the impression persisted that crooked northern merchants were hoodwinking virtuous southern farmers. It didn't help that many New Yorkers in Hamilton's own social circle . . . had accumulated sizable positions in government debt.[3]

Hamilton's "assumption bill" passed Congress, but only after a deal was struck with Thomas Jefferson and James Madison. In exchange for their support, Hamilton agreed that the new national capital would be built in the South. It is also worth noting that, as John Steele Gordon writes in *An Empire of Wealth*, Hamilton's program was an immediate success and the new United States bonds sold out in weeks. "In 1789, the United States had been a financial basket case, its obligations unsalable, its ability to borrow nil," says Gordon. "By 1794, it had the highest credit rating in Europe. . . . The ability of the federal government to borrow huge sums of money at affordable rates in times of emergency—such as during the Civil War and the Great Depression—has been an immense national asset. In large measure we owe that ability to Alexander Hamilton's policies that were put in place at the dawn of the Republic. It is no small legacy."[4]

But many Americans of the day, especially southerners, lacked that historical long view. The Hamilton plan contributed to a growing antagonism and sectional mistrust that would fester dangerously between the regions over the next few decades.

And that's why, to many in the South and West, Burr's role in that "famous affair of honor" on July 11, 1804, would have been cause for admiration. Dueling was no disgrace but a badge of honor, and plenty of men carried the scars to prove it. Some, like the future president, Andrew Jackson, had killed men in duels. And for these reasons, along with the fact that Aaron Burr had a reputation as a bona fide hero of the Revolution, the former vice president still possessed a great many friends and admirers.

Mrs. Hinson told Perkins that the men were headed for Pensacola, Florida, then in Spanish territory. Paddling a canoe down the Mobile River, the determined Perkins reached Fort Stoddert, an army garrison near the border between the American-controlled Mississippi territory and Spanish Florida. Perkins then rode toward Pensacola after the Burr party with an officer, Lieutenant Edmund P. Gaines, accompanied by a detachment of mounted soldiers. Eventually overtaking the three riders, Gaines asked the mystery man if he was "Colonel Burr." With a brief protest that the soldier had no right to arrest him, but otherwise without incident, the former vice president, at that moment the most wanted man in America, was placed under arrest. The president of the United States had publicly accused Burr of treason and declared him guilty—without evidence or a trial, jury, or judge. To Jefferson, the presumption of innocence did not apply to Burr. If convicted, Aaron Burr could hang.

Edmund Gaines would go on to a distinguished army career during the War of 1812 and then continue the long fight against the Seminoles in Florida. (Gainesville, Florida, is named in his honor.) He now took his prisoner back to Fort Stoddert, where the fifty-one-year-old New Yorker soon charmed members of the Gaines family, including Lieutenant Gaines's brother, George S. Gaines, a successful Indian trader. Fearing that Burr would attempt to escape with the help of sympathetic locals, Gaines decided to send him back to Washington.

Nearly legendary in his day for his wit and charm, Aaron Burr had already worked his magic in this wilderness outpost that would eventually be part of the state of Alabama. Refined, educated, debonair—a man of the world who had rubbed shoulders with Washington, Jefferson, and Madison—Burr must have seemed a legend come to life, a man from another world, to these people living on the mostly unsettled edge of America. When he departed from the fort under an armed guard of eight men, to begin the journey back to Washington, Burr had clearly left an indelible mark on some of his acquaintances. "Some women wept, and a mother-to-be later named her child Aaron Burr; neither age nor misfortune had diminished Burr's gift for making friends."[5]

On March 5, 1807, the captive Burr set out on the long trip back to Washington, escorted by Perkins, his friend and fellow attorney Thomas Malone, and six soldiers from the garrison; these eight escorts had already agreed to divide the $2,000 reward for his capture. Working their way through the rough backwoods of Alabama and Georgia, they would have to cover some 1,100 miles of swamp and forest, still home to thousands of Native Americans, who then existed in a very uneasy peace with the growing numbers of white settlers moving into the territory.

At night, Burr was given the single tent the group carried. Perkins was constantly on the alert, not only for hostile Indians but also for an attempt at escape. Burr was known to have many influential friends and allies who might try to free the famous politician. His son-in-law, Joseph Alston, lived in South Carolina, and the Alstons were a powerful force in that state. A wealthy planter, who had served in South Carolina's House of Representatives, Alston had married Theodosia, Burr's daughter, in 1801.*

When Burr and his escort finally reached Chester, South Carolina, Burr made a small and somewhat desperate stab at escape. Jumping from his horse, he shouted to some locals, imploring them to summon a judge. He reportedly declared, sounding more like a lawyer than a dangerous man plotting a coup, "I am Aaron Burr, under military arrest. I claim the protection of the civil authorities."

Perkins, surely thinking that his share of the reward money was about to disappear, grabbed Burr and wrestled the smaller man back into his saddle. His friend Malone slapped the horse's flanks, and the entire party galloped away. Throughout the journey, Burr had maintained a stoic demeanor. But now, with a last chance at freedom lost, he broke. According to the accounts of his guards, Burr wept. Some of his captors joined him.[6]

A little later, Perkins hired a small coach to carry Burr the rest of the way to his date with justice. It would be far more difficult for Burr to bolt from a closed carriage, Perkins reasoned, than from horseback. Initially heading for the nation's capital, the party learned that Burr would be tried instead in Richmond, Virginia's state capital. They arrived in Richmond on March 26, the closed carriage accompanied by "eight dusty riders."[7]

*In a curious bit of American lore, Theodosia Burr and Alston are said to be the first couple to have honeymooned in Niagara Falls.

Word of Burr's approach and the rumored details of his capture had begun to appear in the national press, which by now had turned completely against Burr. Some of the newspapers reported that Burr arrived in Virginia wearing the outfit in which he had been captured. Others speculated that he was wearing a disguise, which he had planned to use when he invaded Mexico. Despite the confusion over his outfit, it was clear that the dapper Burr had taken a precipitous fall from his previous heights.

In fact, it would be difficult to imagine a greater fall from grace in American history than Aaron Burr's at this moment in his life. Perhaps the nearest comparison might be Benedict Arnold, betrayer of the Revolutionary cause, and, ironically, Aaron Burr's first commander during the War for Independence. But Arnold, although a high-ranking officer and an intimate of George Washington, never reached the heights that Burr had visited. And Arnold never had to suffer the consequences of his betrayal. He lived out his years in Canada and London, with considerable success as a shipping merchant. For Burr, it would be a very different story.

An extremely deft politician, Aaron Burr had skillfully maneuvered through the bellwether presidential election of 1800, becoming an architect of Thomas Jefferson's Democratic-Republican* challenge to the Federalist Party of President John Adams and the former secretary of the treasury, Alexander Hamilton. A power in

*Often shortened to "Republican" at the time, this party was actually the forerunner of the modern Democratic Party, the name taken in 1828 during the era of Andrew Jackson. The first Democratic national convention was held in 1832.

New York politics, Burr was instrumental in bringing this Elector-rich state to Jefferson's column as his ostensible running mate. But through a quirk of the process in America's still evolving presidential election system, Jefferson and Burr finished in a tie. The election would eventually be settled in the House of Representatives, where each state would have a single vote. That tie in the voting in 1800 produced America's first serious political crisis since the Constitution had been ratified.

The tale of Aaron Burr's rise and fall is a powerful corrective to the familiar, trumped-up view of the "Revolutionary generation" as a knighted and dignified "band of brothers," a colonial-era Kiwanis Club of glad-handing gentlemen united by a singular cause and ideology. For all their extraordinary accomplishments in creating an independent America out of thirteen disparate colonies in 1776 and then forming "a more perfect Union" in the summer of 1787, the men called Founding Fathers or Framers were human beings.

And they were politicians, with all the baggage that the word implies. They were ruled by many of the same forces that rule modern American politics—greed; self-interest; regional and commercial interests. They were quite practiced at the "politics of personal destruction." And of course, they were capable of grave personal misconduct. They had affairs. They kept slaves. They dissembled and brokered deals. Perhaps these very human failings and contradictions make their achievements all the more remarkable. Some of their names and faces grace America's currency. Others have fallen into ignominy. And some were turned into villains by historians in-tent upon fashioning a comfortable American mythology. Deserving or not, Burr made a perfect candidate for villainy. And that is how he has been depicted for much of the past 200 years.

Born on February 6, 1756, in Newark, in what was then colonial New Jersey, the man whose behavior would later scandalize New York in the early days of the new nation was welcomed into the world of two of the most influential, important, and conservative ministers in America. In fact, Aaron Burr was supposed to follow in their paths. His father, also named Aaron Burr, was a respected Presbyterian preacher who helped found, and then became the second president of, the College of New Jersey (later renamed Princeton University). His mother, Esther Edwards Burr, was the daughter of the famous and highly influential American preacher Jonathan Edwards, the noted Calvinist theologian who was among the leaders of the religious movement known as the Great Awakening.

The first of several waves of fundamental, orthodox Protestantism that periodically swept over America, the Great Awakening of the 1730s and 1740s spread like wildfire from New England into the other colonies. It was largely a response to what colonial American clergymen viewed as a slackening of religious values in the increasingly prosperous young America. As the colonial economies improved, many Calvinist ministers watched with dismay as their congregations, once fully devoted to the Sabbath practices of their Pilgrim and Puritan forebears, turned to such earthly pursuits as real estate speculation, slave trading, the rum business, and other equally profitable enterprises. This initial Great Awakening also came as Enlightenment ideas about reason and science were shaking the ancient traditions of religious philosophy. A decade later, those Enlightenment ideas would burst through in the form of deism and a new belief in the rights of man, which contributed powerfully to the Revolutionary mood in an America on the road to independence.

One of the intense sparks behind this dynamic revival of old-fashioned fire-and-brimstone Calvinism was the arrival in America, in 1739, of an Anglican preacher, George Whitefield, whose reputation as an orator brought thousands to his outdoor meetings, large-scale revivals that might be likened to the modern "crusades" of Billy Graham. In Philadelphia, then America's largest city, Whitefield's "born again" evangelism drew 6,000 listeners, nearly half the city's population of about 13,000. In his powerful, emotionally charged sermons, Whitefield chastised his listeners and then offered them the promise of salvation. Surprisingly, Whitefield even won the admiration of Benjamin Franklin, who eventually published forty-five of the sermons in his newspapers, eight of them on the front page of the weekly *Gazette*.

For Franklin, himself a deist and Freemason whose appetite for orthodox religion was meager, Whitefield's appeal lay in his powerful admonitions to do good works. America's great apostle of philanthropy, Franklin was singularly impressed by the fact that Whitefield raised more money than any other cleric for orphanages, schools, libraries, and almshouses across Europe and America.[8] Whitefield's Christian principles, however, left plenty of room for African slavery, which he advocated and which Franklin himself only gradually came to oppose later in his career.

Franklin's esteem made Whitefield what some have called America's "first celebrity." There is a notion that the saturation tactics of Madison Avenue media and marketing are a contemporary American invention, but Whitefield pioneered the development of "multiplatform" marketing strategies. As the historian Larry Witham points out, "He achieved this celebrity by a canny use of letters, news accounts, advertising, advance teams, strategic controversies, and dramatically staged events—in a word, the first mass-media campaign in America."[9]

As Ecclesiastes tells us, "There is nothing new under the sun."

The other great force occupying an American pulpit at this time was Jonathan Edwards, a Congregational preacher born in Connecticut in 1703. Having entered Yale at age thirteen, Edwards was head tutor at the college by age twenty. He later took the pulpit at the Congregational Church in Northampton, Massachusetts, where his grandfather had been preacher. A Calvinist of the old school, Edwards believed that grace or salvation was given only to the "elect," those predestined at birth. But he also gave a nod to the thought of the day by acknowledging science and the philosophy of men like John Locke and Isaac Newton. As he sought to separate the wheat from the chaff, the elect from the hopeless sinners, his preaching had an enormous impact on colonial America.

Unlike the theatrical Whitefield, whose voice was dramatic and booming, Edwards spoke blandly and without gestures. But his words sent his audiences into paroxysms of wailing and horror. Perhaps his most famous sermon, "Sinners in the Hands of an Angry God," conveys some of the sense of dread his congregations must have felt:

The God that holds you over the pit of hell, much as one holds a spider or some loathsome creature over the fire, abhors you, and is dreadfully provoked; his wrath towards you burns like fire; he looks upon you as worthy of nothing else but to be cast into fire; he is of purer eyes than to bear to have you in his sight; you are ten thousand times so abominable in his eyes as the most hateful and venomous serpent is in ours.

This first Great Awakening provoked a controversy over theology that divided many American Protestant denominations of the time into groups called the "New Lights"—followers of Edwards and other Great Awakening fundamentalists—and "Old Lights," who opposed the emotionalism of the revival movement. The schism eventually split some of these denominations, spawned new ones, and created divisions within some of the prestigious colleges, such as Harvard and Yale, where the faculty and student body were soon choosing sides. Jonathan Edwards himself wore out his welcome, alienating his own flock. Turned out by his congregation, Edwards left Northampton in 1750, and set out to become a missionary to the Indians.

In the meantime, his son-in-law Aaron Burr (Senior) and his fellow clergyman Jonathan Dickinson, another "New Light" leader, had departed from Yale in a dispute over these differences between "Old Light" and "New Light" and founded the College of New Jersey in 1746. The school was first established in Dickinson's home in Elizabethtown, New Jersey, and Dickinson was elected as its first president. But when he died less than six months later, a victim of smallpox, Burr replaced him as president. Under Burr's leadership, the College of New Jersey grew to much greater prominence. Moving the school to its permanent home at Princeton, New Jersey, Burr supervised construction of Nassau Hall, the largest building in British North America at the time of its completion in 1756.*

But the world of the Edwards-Burr family soon came crashing down in an astonishing series of tragedies that unfolded like one of Charles Dickens's grimmer novels and serve as reminders of the frailty of life in colonial America. In September 1757, Aaron Burr Senior died after a three-week illness. His father-in-law, Jonathan Edwards, emerged from the wilderness and moved to Princeton to replace Burr as head of the college. But within six months, Edwards himself succumbed to smallpox, the great killer of colonial America. In March 1758, young Aaron Burr's mother, Esther Edwards Burr, also fell ill, although she and her two children had been inoculated against the disease. She died on April 7, 1758, a few months after her son Aaron's second birthday. The children's grandmother, Sarah Edwards, Esther's mother, then came to Princeton to collect the orphaned Aaron and his older sister, Sarah (also called Sally). But Sarah Edwards fell ill, too, of dysentery, and was dead by October.

Suddenly, young Aaron Burr and his sister Sarah were without family. For two years, the children were raised in the Philadelphia

*A grandson of Edwards's, Timothy Dwight, would become the president of Yale in 1795. An influential theologian, Dwight was also a prominent Federalist Party leader. He was at the forefront of what would become known as the "Second Great Awakening" at the beginning of the nineteenth century.

home of a family friend, Dr. William Shippen, later one of the founders of the University of Pennsylvania Medical School.* In 1760, the two children were taken into the Sturbridge, Massachusetts home of their uncle, Timothy Edwards, their mother's younger brother, who was a minister like his father, Jonathan. Young Aaron Burr was raised in the midst of an extended family of aunts, uncles, and other orphaned cousins.

Timothy Edwards eventually moved the entire clan to Elizabethtown, New Jersey, and gave up the "cloth" for the law, becoming a successful attorney.† Like many children of the time, Aaron Burr was expected, from childhood, to follow in his father's footsteps. But chafing under the discipline of his uncle, young Aaron bolted from home several times, once signing on as a cabin boy on a merchant sailing ship. He was collected before the ship sailed.[10]

When he was eleven, Aaron Burr's application was submitted to the College of New Jersey, initially established to train young men for the ministry, just as Harvard and Yale were. The college, like many colonial-era American "colleges," was more like a prep school than a modern university. Despite his familial connections to the school, Burr was initially rejected as too young. He was tutored at home by a Princeton graduate, Tapping Reeve, and reapplied two

*In another delicious historical connection, one of Shippen's young relatives, Peggy Shippen, would later become the second Mrs. Benedict Arnold, the teenage wife who was totally complicit in Arnold's betrayal of the patriot cause.

†Uncle Timothy Edwards later became a military contractor during the Revolution and a successful land speculator after the war.

years later. This time Burr was admitted as a sophomore, four years younger than most of his classmates. Owing to his youth, his much-admired father, and his diminutive stature—Burr was about five feet six inches when fully grown—his Princeton classmates dubbed him "Little Burr." Among those fellow students were Jonathan Dayton, a childhood friend and future delegate to the Constitutional Convention; Light Horse Harry Lee, future Revolutionary War hero and father of Robert E. Lee; and Luther Martin, another delegate to the Constitutional Convention who would become a famous trial lawyer—and defend Aaron Burr.

These were extraordinary times in colonial America, with the winds of change blowing, especially in places like Princeton. The college was by then under the direction of John Witherspoon, an influential Presbyterian clergyman who had arrived from Scotland and was destined to have a profound impact on the Revolutionary generation. As the historian Arthur Herman describes Witherspoon:

> *He intended to make Princeton not only the best college in the colonies, but in the entire British world. . . . Witherspoon saw education not as a form of indoctrination, or of reinforcing a religious orthodoxy, but as a broadening and deepening of the mind and spirit—and the idea of freedom was fundamental to that process. "Govern, govern, govern always," he told his faculty and tutors, "but beware of governing too much. Convince your pupils . . . that you wish to see them happy."*[11]

To young Aaron Burr, such talk of "happiness" was surely far removed from the dangling, barbecued spiders of his departed grandfather, Jonathan Edwards.

Witherspoon encouraged the study of modern philosophers, including Locke and Hume, as well as the classics. He transformed Princeton's student clubs into venues for rigorous intellectual discussion, and two of his best and most enthusiastic students were Aaron Burr and a Virginian, James Madison, Burr's senior by five years. Increasingly drawn to the burning political questions of the day, Witherspoon later wrote an influential essay, "Thoughts on American Liberty." And in May 1775, after the battles of Lexington and Concord, as the Continental Congress began the process of declaring America's independence, Witherspoon preached a sermon later published as "The Dominion of Providence over the Passions of Men." In it, Witherspoon invoked a divine blessing on the righteousness of the American cause. In June 1776, he joined the New Jersey delegation to Philadelphia and the Continental Congress. He was there to help pass the Declaration of Independence, which he later signed.

It was against this backdrop, with growing talk of independence, the rights of man, and resistance to the crown, that sixteen-year-old Aaron Burr graduated from Princeton in 1772. Expected to follow his father and grandfather into a pulpit, he briefly trained for the ministry in Connecticut before writing to inform his uncle that he preferred to pursue a career in law. When his sister Sally married Tapping Reeve—their former tutor, now an eminent legal scholar who would open one of the first law schools in America—Burr moved into the couple's home in Litchfield, Connecticut, and began to study law.

But the world was about to be turned upside down. After the fighting broke out in Concord and Lexington in April 1775, Aaron Burr joined the thousands of other idealistic, adventurous, and starry-eyed young men who trooped off to Cambridge, on the outskirts of Boston, where some 20,000 very undisciplined Americans had collected to become the ragtag Continental Army. By the time Burr arrived, they were under the command of General George Washington, to whom Congress had presented the unenviable task of creating an American army. Washington's mission was to take this motley band of mostly undisciplined, untested vagabonds, backwoodsmen, farm boys, and out-of-work laborers and train them to fight the most powerful force on the face of the earth, the combined British army and navy.

At age nineteen, with no military experience but considerable enthusiasm, the preacher's son turned law student learned that an expedition was being formed to invade Canada by traversing the wilderness of Maine, just as another expedition to be led by General Philip Schuyler moved on Canada from upstate New York. George Washington hoped to win French-Canadian support for American independence and perhaps create a "fourteenth state" to America's north. Washington gave command of the mission to the man who brought him the idea—Benedict Arnold.

Largely responsible for the successful assault on Fort Ticonderoga on Lake Champlain in June 1775 that had been a boon to the patriot cause, Arnold believed that a two-pronged assault on Montreal and Quebec could possibly force the British to abandon

Canada and bring them to the peace table. Washington agreed and, along with his blessing, gave Arnold a letter to the people of Canada hoping to inspire a little revolutionary fire, and orders to respect the religion of the predominantly Catholic French Canadians. Washington also sent along some of the best troops he had, the handpicked riflemen led by Daniel Morgan, Daniel Boone's cousin, and another hero of the Revolution.

Burr and his cousin Matthias Ogden, who had also grown up in the extended Edwards household, were able to join the expedition. Arnold welcomed both young men, perhaps because of their family connections. In Revolutionary America, what you knew was still considerably less significant than who you were.

After a brief pilgrimage to visit the crypt of George Whitefield in nearby Newburyport, Massachusetts, Arnold's battalion set off in October to surprise the British at Quebec. The 350-mile trek through the backwoods of Maine proved disastrous, one of the most horrendous in American military history. Rain-flooded rivers swept away the company's hastily and poorly built boats, and with them many supplies. In short order, Arnold's men were reduced to eating dogs. Their boots were worn away, leaving many of the men barefoot in the Maine winter. Matthias Ogden wrapped his feet in a flour sack. Some men ate their own shoe leather. Starvation and, with it, rampant diseases like smallpox, the ever-present Revolutionary-era killer, took a grievous toll on Arnold's force.

After the six-week march, more than one-third of the 1,100 men who had left Boston with Arnold were sick, dead, or missing. When Arnold's emaciated and exhausted troops arrived on the outskirts of Quebec in November 1775, they were a far cry from a well-prepared strike force capable of taking one of the British strongholds in Canada. But as Thomas Fleming records, "Arnold's half-starved men amazed the British when they emerged from the Maine wilderness before Quebec. . . . The French were ready to open the gates to Arnold and his 675 ragged scarecrows. Many of the habitants (French Canadians) thought only beings with miraculous powers could have survived the privations of their 350-mile march."[12]

But a sudden deathblow to British Canada was not to be. Quickly reinforcing the fortress city, the British forced any Frenchmen sympathetic to the American cause out of Quebec, and Arnold was left to settle in and await reinforcements.

Young Aaron Burr had endured the dreadful forced march with flying colors. Upon arriving at Quebec, the very green but very enthusiastic officer was dispatched to meet with General Richard Montgomery, who had captured Montreal and was a day's march away. An Irish-born veteran of the British army, Richard Montgomery had fought in the French and Indian War (or Seven Years' War) and was one of the most battle-tested commanders in the relatively untested Continental Army. After marrying into New York's influential Livingston family, Montgomery had joined the American cause; he was leading the second front of the assault on Canada in place of General Philip Schuyler, who had fallen ill. Having captured Montreal with relative ease, Montgomery was planning to link his forces with Arnold's troops for the assault on Quebec. In dispatching Burr to meet Montgomery, Arnold's note to Montgomery recommended the teenage officer: "He is a young gentleman of much life and activity, and has acted with great spirit and resolution on our fatiguing march."[13]

Montgomery took an instant liking to Burr, promoting him to captain, and choosing him as an aide-de-camp, so Burr was at Montgomery's side on the fateful night of December 31, 1775. In a daring, perhaps reckless, attempt to take Quebec, the combined forces of Montgomery and Arnold assaulted the fortress city in a howling snowstorm. Using ladders to scale the city walls, some Americans led by Daniel Morgan actually breached Quebec's walls. They were to be augmented by assaults from two other directions: one led by Montgomery, the other by Arnold. But through a combination of bad luck, bad weather, and superior British firepower, the tide quickly turned.

In the hand-to-hand fighting within the city walls, Daniel Morgan was captured, along with more than 300 Americans. Wounded in the leg, Benedict Arnold was forced to withdraw. More disastrously for the American cause, General Montgomery was killed by a burst of grapeshot fired by a Loyalist who had left Boston for Canada. When Montgomery fell, some of the American fighters took to their heels.

Aaron Burr was present when Montgomery died, and his actions later led to controversy. In the widely accepted version of the day, Burr had tried to rally Montgomery's troops, who retreated instead. Then, Burr attempted to lift the general, who had reportedly died in his arms, and carry the body from the field. This scene would be depicted thirteen years later in John Trumbull's painting of 1758, *The Death of Montgomery at Quebec*, which became an icon for patriotic Americans, with copies hanging in schoolrooms around the country. That painting alone was one reason why many Americans, including some of his later captors, deemed Aaron Burr a hero. The

British recovered Montgomery's body, and he was buried with full military honors.

In later years, as Burr's reputation came under attack during the political and personal controversies that dogged him, the truth of what happened that night in Quebec also got buried. Burr's critics attempted to blacken his name with whispers that Burr had retreated from Quebec.

But there was no question that this stunning defeat, as the epochal year of 1776 dawned, was critical. Had it succeeded, the siege of Quebec might have changed the course of the war. Instead, Benedict Arnold withdrew his men and took up positions around Quebec, awaiting reinforcements. After a bitter winter of disease and starvation, Arnold finally retreated from Canada in the spring of 1776. In recognition of his service, Aaron Burr, now twenty, was promoted to major, and in June 1776 he joined the staff of General George Washington in New York City. Washington had moved the army to New York after the British evacuated Boston a few months earlier. Washington's staff was occupying a mansion and large property known as Richmond Hill, in what is now Greenwich Village. (The building, which no longer stands, was located at what is now the intersection of Charlton and Varick streets, and was later purchased by Aaron Burr as his New York residence.)

Confronting the British in New York proved one of Washington's most disastrous decisions. Even as the Continental Congress was voting for independence on July 2, 1776, and adopting Thomas Jefferson's Declaration of Independence two days later, the British were moving one of the greatest armadas ever seen into New York harbor. Over the course of the next two months, more than 350

British warships arrived in New York, delivering some 30,000 well-trained and well-equipped British soldiers and German mercenaries. Washington had about half that number—most of them ill equipped and inexperienced.

Later accounts of the relationship between Washington and Burr would paint a picture of immediate mutual disdain. But those reports were probably colored by events long after the war, when Burr's reputation began to take a beating. Eager for a more active role, Burr was assigned to the staff of General Israel Putnam before even encountering Washington. One of the heroes of the Battle of Bunker Hill, "Old Put" was a Connecticut Yankee who had fought with distinction in the French and Indian War, and once survived a near roasting at the hands of his Indian captors. According to Revolutionary lore, he left his plow at the first word of fighting at Lexington and Concord, and drove his own cattle to Cambridge to feed the troops. It was Putnam who supposedly uttered the famous—but possibly mythologized—"Don't fire until you see the whites of their eyes" at the Battle of Bunker Hill.

When the battle of New York finally came in mid-September 1776, it proved a rout. Washington was fortunate to get most of his army out of New York to fight another day. Burr helped oversee the successful retreat of troops from lower Manhattan to Harlem. And on September 16, in the midst of an otherwise general disaster for the Americans, Burr won a battle in Harlem Heights, further burnishing his reputation as a decisive, courageous leader.

Washington's reputation, on the other hand, began to suffer the first of many setbacks after the debacle in New York. The near-destruction of his army and the heavy losses of men and matériel he had absorbed set off the first round of whispers that he should be replaced. The whispers grew to a movement as several congressmen wondered if they had chosen the right man in George Washington. And it was a sort of dividing line in terms of loyalty to the commander. You were either with Washington or against him. As Burr's biographer Nancy Isenberg recounts, "This was the start of an obsession among Washington's staff, talk of 'cabals' or secret plots, which in turn fueled the 'party business' that Burr assiduously wished to avoid. In his yearlong tenure as Putnam's aide, Burr clearly saw the worst side of Washington, yet the historical record gives no evidence that he ever reproached the embattled commander."[14]

Only later, after the war, did Washington view Aaron Burr as a member of the "opposition," and it was only then that Burr's name became attached to these anti-Washington conspiracies.

Burr was initially passed over for promotion, and this peeved him—much as it did Benedict Arnold when he was similarly passed over, with much more disastrous results. But eventually Burr was given a command of his own and served with distinction, suffering through the Valley Forge winter of 1777–1778 as a commander who won high praise for the discipline of his men. By 1779, however, Burr's health—worsened by years of exposure and poor diet—did him in. Bedeviled by debilitating migraine headaches, and exhausted by the internal politics of the army that have frustrated many a soldier, he tendered his resignation to George Washington in March 1779. At age twenty-three, Burr was a retired soldier. During the war years, he again won praise as he took the field in defense of Connecticut during some British attacks. But as the focus of the fighting moved south, Aaron Burr was far removed from the eventual American victory at Yorktown in October 1781.

Having resumed his legal studies, Burr passed the New York state bar exam in April 1782. A few months later, the young attorney's life was again greatly altered—by marriage. On July 2, 1782, Burr wed Theodosia Bartow Prevost, the widow of James Marcus Prevost, a British army officer who had succumbed to yellow fever while posted in the West Indies during the war. Ten years Aaron Burr's senior and the mother of five children, the French-speaking Theodosia was cultivated, articulate, and alluring. She apparently attracted both young Patriots and British officers.

Burr and Theodosia had met when she made her expansive New Jersey estate near Paramus available to George Washington as a headquarters in July 1778. Despite being the daughter of one British officer and having married another, Theodosia moved easily between the worlds of Patriots and Loyalists. But her instincts seemed to favor the American cause. One hint of her political leanings was the fact that she had named her estate the Hermitage, after the French philosopher Jean-Jacques Rousseau's celebrated cottage—though Theodosia's Hermitage was anything but a "cottage." With ninety-eight acres and two elegant homes, the Hermitage offered a taste of polite European aristocratic society set down in the midst of a war zone.

Burr first encountered Theodosia while serving as a military escort for three prominent Loyalists in August 1778. What began as friendship changed with Burr's offer to assist Theodosia in dealing with some of her legal matters, including protecting the Hermitage—which was officially viewed as "Loyalist property"—

from being appropriated by American authorities. Their friendship soon became much more. By the time Theodosia's husband died in the West Indies in 1781, the rumors and wagging tongues of the day seem to have been borne out.

Burr's relationship with Theodosia was unusual if not extraordinary for its time. Despite his reputation, deserved or not, as a licentious womanizer, Aaron Burr might be called a "proto-feminist." Not only was Burr an early abolitionist in a state that had more slaves than any other in the North; he also proposed legislation in the New York state assembly that would give women the vote. He believed in the equality of the sexes and treated his wife accordingly. "Burr distinctly pursued a marriage based on a very modern idea of friendship between the sexes," writes Isenberg. "He found such advocacy in the writings of John Witherspoon, president of his alma mater, Princeton, . . . and in Mary Wollstonecraft's *Vindication of the Rights of Woman*." This feminist manifesto of 1792 argued that women were not inferior to men but were simply denied opportunity and education. Burr was "practicing its egalitarian marital principles ten years before its publication."[15]

In addition to Theodosia's children, the couple had two daughters who survived birth. Theodosia, named after her mother, was born in 1783. A second daughter, Sally, died at age three; and three more stillbirths followed—another harsh reminder of the infant mortality rate in early America.

Burr and his wife were devoted to their only surviving daughter, and she received an education that few girls of her day could dream of. Clearly a prodigy, the young Theodosia studied mathematics, geography, Latin, Greek, and French. "Believing that she was the equal of any man, Burr educated her as he would have a son," Joseph Wheelan writes. "His advocacy of women's education was rare in an age when girls were taught little beyond simple reading and writing. As he once declared to his wife, Burr wished to 'convince the world what neither sex appear to believe—that women have souls.' Burr believed that women's education was of paramount importance because children received their first impressions almost exclusively from their mother, the 'repositories of all the moral virtues' that went into the making of men of 'excellence.'"[16]

Burr began his political career in earnest in 1782, when he was elected to the New York state assembly. When the war finally ended with the Peace of Paris in 1783, he moved from Albany to New York City. His extended family, including his wife, her children, and Theodosia, moved through a succession of homes as Burr's legal practice and reputation flourished. Joining the group of young Patriot lawyers establishing themselves in New York, Burr had begun to build a reputation as a brilliant trial lawyer known for his "cogency, precision and quiet sarcasm."[17] His career in ascendancy, Burr was able, by 1791, to purchase the 6,000-square-foot house, Richmond Hill, that had been Washington's staff office and John Adams's home when Adams was vice president during the brief period that New York was the nation's capital. The estate included twenty-six acres of grounds that sloped down to the Hudson River. By then Burr, who had purchased the Hermitage from Theodosia's family, had begun to sell off the estate, to raise money.

At the same time, Burr began to exercise considerable influence in New York politics. In 1789, at thirty-three, he had become New York's attorney general, and two years later he was elected to the United States Senate from New York by the New York state legislature.* Burr defeated the incumbent, the Revolutionary War general Philip Schuyler, who also happened to be the father-in-law of Alexander Hamilton, another lawyer of the Patriot generation.

Burr continued his progress through New York state politics just as the lines were becoming clearer between the two emerging national political parties. On one side were the Federalists, whose tenet was a strong central government. President John Adams and Alexander Hamilton led this party. However, they neither agreed on every policy nor especially liked each other.

On the other side were the Democratic-Republicans (or "Republicans"), whose philosophy was inclined toward letting the states retain more power and limiting the powers of the federal government. In foreign affairs, the Federalists tended to view Great Britain as an ally, while the Republicans, including Thomas Jefferson, preferred France—England's chief continental enemy—as America's truest ally.

Even in his Senate victory over Schuyler, Burr had remained aloof from either party. But once in the Senate, he was drawn toward the emerging Democratic Republican Party of Thomas Jefferson. When the national government later moved to Philadelphia during the Washington administration, Burr and Jefferson roomed at the boardinghouse of a Mrs. Payne, where Burr met the landlady's attractive daughter Dolley, who had been widowed when yellow

*Under the original Constitution, United States senators were elected by state legislatures. In April 1913, the Seventeenth Amendment, which provided for direct election of senators by popular vote, was ratified.

fever struck Philadelphia in 1793. In 1794, Burr played matchmaker, introducing Dolley Payne to his former schoolmate at Princeton, James "Jemmy" Madison, then a forty-four-year-old bachelor, and the two were married that year.

But Burr's own domestic happiness would soon be shattered. In May 1794, while Burr was attending Senate sessions in Philadelphia, he was stunned by the death of his wife Theodosia at age forty-eight. Although she had been ill for years—most likely with cancer—and was being treated with laudanum, an opiate, her death was unexpected and crushing. By all accounts an extraordinarily progressive, intelligent, well-read woman, she was not only Burr's very loving wife but an excellent political adviser as well. During the war, she had navigated through the sharp rocks between the Loyalists and Patriots and would have been useful as Burr dealt with his political storms. Burr would later say that the loss, "dealt me more pain than all sorrows combined."

When Washington chose not to run again after his second term, Vice President John Adams succeeded the "father of the country" as the second president. In that first contested election of 1796, the political lines were already being drawn more sharply. The "factions" or party politics that Washington had warned against and scrupulously tried to avoid were clearly emerging. And Burr's image and reputation as a power broker were also being more sharply defined, fairly or not.

By this time, Alexander Hamilton had suffered his own fall from grace. His downfall came from the political press he had helped create. Americans had become increasingly literate, and the press was becoming increasingly powerful, a reality that was reinforced in the new republic as America witnessed an explosion of partisan newspapers and political journals and pamphlets. Benjamin Franklin Bache, Franklin's grandson, published a paper, known popularly as the *Aurora*, which became an outspoken organ of the anti-Federalist cause. Sponsored by Alexander Hamilton, the *Gazette of the United States*, published by John Fenno, was a powerful voice of Federalism.

But Hamilton would himself be attacked in print by the early republic's equivalent of the Drudge Report. Working as a political propagandist for Jefferson's party, the Scottish-born scandalmonger and pamphleteer James Callender wrote *History of 1796*, a pamphlet in which he exposed Hamilton's adulterous affair with a woman named Maria Reynolds. When word got out that Maria Reynolds's husband was blackmailing him, Hamilton was forced to resign as secretary of the treasury early in 1795. Hamilton would not hold public office again. With incidents like these to illustrate the poisonous power of the pen, politicians like Jefferson, Hamilton, and Burr knew full well the impact of a partisan press, and the newspapers would play an increasing role as Aaron Burr's drama unfolded.

In a sublimely ironic footnote to this story, Hamilton blamed the revelation of the affair on James Madison, once his closest colleague in framing the Constitution. Madison was now increasingly linked with his fellow Virginian Thomas Jefferson's Democratic Republicans. As Hamilton and Madison contemplated a duel, none other than Aaron Burr stepped in to conciliate and play peacemaker.

Whatever the previous relationship between Burr and Hamilton, by the time Hamilton returned to New York, he increasingly came to despise Burr. What is more, Hamilton still had Washington's ear. It may have been Hamilton who derailed Burr's appointment as minister to France. And when a possible war, the "Quasi-War," with France was brewing in 1798, George Washington was appointed commander of the U.S. forces and quashed Burr's application for a commission. According to President Adams's later recollections, Washington had told him, "By all that I have known and heard, Colonel Burr is a brave and able officer, but the question is whether he has not equal talents at intrigue."[18]

Bored by a fairly inactive Senate, Burr returned to New York, where he was elected to the New York state assembly, serving from 1798 through 1801. Associating loosely with the Republicans, Burr still maintained contacts with moderate Federalist allies, including his old friend, now a senator, Jonathan Dayton of New Jersey. Still on Burr's legislative agenda was a plan to emancipate New York's slaves and protect its free blacks, not a popular platform in the 1790s.

But he was also quickly moving to assert control over political events in New York. Burr worked through the Tammany Society, later to become the infamous Tammany Hall, which he converted from a social club into one of the first organized urban political machines. At around the same time, Burr demonstrated his skills at manipulating the process when he used a bill in the state legislature that established a water utility to help start the Bank of the Manhattan Company (which in later years evolved into the Chase Manhattan Bank, now JPMorgan Chase.) John Steele Gordon explains in his history of American economic power, *An Empire of Wealth*: "At the turn of the century, obtaining a bank charter required an act of the state legislature. This of course injected a powerful element of politics into the process and invited what today would be called corruption but then was regarded as business as usual. Hamilton's political enemy . . . Aaron Burr was able to create a bank by sneaking a clause into a charter for a company called the Manhattan Company, to provide clean water to New York City. The innocuous-looking clause allowed the company to invest surplus capital in any lawful enterprise. Within six months of the company's creation, and long before it had laid a single section of water pipe, the company opened a bank."[19] Hamilton had already founded the Bank of New York in 1784, and now he and Burr were competing in the world of finance as well.

Their increasingly heated rivalry came to a head in the election of 1800, when Burr emerged as a "kingmaker," with unexpected results. Burr had now aligned himself firmly with his old friend James Madison behind the candidacy of Thomas Jefferson, who had lost the election of 1796 to John Adams. Under the existing electoral process, in which the second-place finisher became vice president, Jefferson had become Adams's very unhappy vice president. The two revolutionaries who had worked together to draft the Declaration of Independence had gone in separate political directions. Adams was an unapologetic Federalist. Jefferson led the Democratic-Republicans. The Framers had not foreseen the evolution of parties, or "factions," so quickly—nor the prospect that two men of such differing views could wind up as president and vice president.

In the spring of 1800, Burr was able to swing the New York state legislature to a Republican majority. He could now deliver the state's rich lode of Electors to Jefferson. Recognizing that power, Burr was placed with Jefferson on the Democratic-Republican presidential ticket in 1800. Unexpectedly, the two men finished in a tie for the presidency, with seventy-three electoral votes each. The election of 1800 would be decided in the House of Representatives, in a special session, which began on Wednesday, February 11, 1801.

The election of 1800 was a crucible in the life of the young republic. There was talk of disunion in the air. There were threats that a Federalist army was being organized to march on Washington and assassinate Jefferson in a bloody coup. Federalists who were certain to lose the presidency spoke of the stalemate allowing Adams to continue as president and of calling a new election. It was a dangerous and uncertain moment with the nation's future at stake as the powers of Europe—Great Britain, France, and Spain, all still holding enormous portions of North America—watched and waited.

That there were intrigues and dealings behind the scenes is small wonder. Some Federalists thought that a Burr presidency would be preferable to seeing Jefferson—the "atheist"—as president. But here is another point on which historians still argue: Did Burr maneuver secretly to become president? Or did he honor his commitment to Jefferson and plan to accept the vice presidency? Many think that Burr and his associates actively campaigned for Federalist votes, attempting to "steal" the office from Jefferson, the "sage of Monticello" and make Burr president.

In her generally admiring biography of Burr, Nancy Isenberg convincingly argues that the idea of a Burr "conspiracy" makes no sense:

> *It is . . . farfetched to suggest that Burr would abandon the Republican Party at this moment, given the decisive role he had undertaken in transforming New York into a Republican state. He would have*

lost his base, the loyal supporters he had acquired over the years, especially through his labors in the state assembly. No politician could maintain his national stature without a strong following in his own state. Does it really make sense that Burr would sacrifice everything he had worked for? He was still only forty-six, and in eight years, he would be in line for the presidency.[20]

One man taking an active interest in the outcome was Alexander Hamilton, who still pulled strings in the Federalist Party. Hamilton disliked both candidates, but he distrusted Burr. He undertook an intense letter-writing campaign designed to undermine his fellow New Yorker.

It took thirty-six ballots before James A. Bayard, a Delaware Federalist, submitted a blank vote, giving Delaware to Jefferson. Bayard's decision, it seems, came after he received a guarantee that Jefferson would strengthen the navy and Federalist officeholders would retain their posts. In other words, horse-trading and patronage were probably more significant than Hamilton's character assassination in giving Jefferson his victory. Other Federalist voters in Vermont and Maryland followed Bayard's lead, and on Tuesday, February 17, 1801, after thirty-six ballots, Thomas Jefferson became America's third president.*

For his part, Jefferson no longer trusted Burr, and the New York power broker who had done more than anyone to ensure Jefferson's

*To avoid repeating such an outcome, the Twelfth Amendment was proposed in 1803 and quickly ratified in 1804. It called for separate elections of the president and vice president.

election was effectively shut out of Jefferson's administration and party politics. Their relationship turned frosty. Excluded from the president's ruling inner circle, Burr was left to preside over the Senate, his constitutionally mandated role as vice president.

Almost from the outset of his first term, it was evident that Jefferson would drop Burr from the Republican ticket in the 1804 reelection campaign. So Burr chose to return to his power base and instead run for governor of New York in the election of April 1804. Burr lost the election and blamed the defeat on a savage personal smear campaign. As early as 1801, a pamphlet attacking Burr's politics and personal behavior had appeared. In it, he was accused of "abandoned profligacy," and there were references to numerous "wretches" he had seduced, many of them since reduced to courtesans; others diseased or dead.[21]

Initially, Burr had ignored these assaults on his character and reputation. But after his defeat in New York, that changed. He began to focus his anger on Alexander Hamilton, the man who he now believed was behind the campaign to destroy him. Their long-simmering feud came to a boil when Hamilton, speaking at a political dinner, announced that he could express a "still more despicable opinion" of Burr. After an account of this incident was published in the *Albany Register*, Burr sought an explanation from Hamilton.

Unsatisfied by Hamilton's reply, Burr, still the sitting vice president, then issued a challenge under the code duello, the formalized rules of dueling. Although dueling was technically illegal in several states, including New York, these affairs of "honor" were often carried out anyway in a scripted dance. Usually, there was a harmless exchange of shots and both participants could walk away from the field having avenged their "good names." But that was not always true, as Hamilton was painfully aware. His eldest son, Philip, had been killed in a duel three years earlier.

On July 11, 1804, the two men and their parties met outside Weehawken, New Jersey (the site is above the Hudson River, not far from where the Lincoln Tunnel's famous Helix now empties into New Jersey). Both men shot. Hamilton missed; Burr's shot entered Hamilton's abdomen above his right hip, piercing his liver and spine. Hamilton was evacuated to Manhattan, where he lay in the house of a friend, receiving visitors until he died in agony the following day.

As with a number of the crucial aspects of Burr's life, what actually happened remains controversial. Almost immediately, Burr was cast as a murderous villain in most of the New York press. In death, Hamilton began to be glorified. Both men clearly fired their pistols, but witnesses could never agree on who shot first. Nor do contemporary historians. Ron Chernow's largely admiring biography *Alexander Hamilton* makes the case that Hamilton did fire first, but missed deliberately, and that Burr fired seconds later, purposefully aiming to kill.[22] On the other hand, Isenberg's account questions the presumption that Hamilton intended to miss. She suggests that there is ample evidence to the contrary.[23] In his account in *Duel*, Fleming notes, "As for who fired first, the two guns went off almost simultaneously. . . . After two hundred years of controversy, including some absurd magazine articles portraying Hamilton as a dishonorable sneak who had secretly set his hair trigger when [a witness] was not looking, it seems simplest to assume everyone was telling the truth as he saw it, with a minor amount of embellishment on both sides."[24]

That duel may never be settled.

What is clear is that Burr was indicted for murder, in both New York and New Jersey. New York later dropped the murder charge, but he was still charged with violating the dueling law, though the "crime" had taken place in New Jersey. And the charge in New Jersey came despite the fact that dueling was legal in the state. But Aaron Burr would never be tried in either state. He had already fled south, to take refuge in the home of a friend.

After Burr eventually returned to Washington, his final act in office was to preside, as the sitting vice president and president of the Senate, over the first impeachment trial in American history, that of Supreme Court Justice Samuel Chase. In a case that was clearly political in nature, Chase—a signer of the Declaration—was brought to trial in the aftermath of a series of rulings from the bench against Jeffersonians, including the sedition trial of the journalist James Callender in 1800. During the Adams presidency, Callender had been convicted under the 1798 Alien and Sedition Acts, one of which made it a crime to publish "false and malicious writings" against the government. As president, Jefferson later pardoned Callender, and Chase was impeached. Burr was then responsible for overseeing the trial of a Federalist justice in an overwhelmingly Republican Senate.

But in spite of Federalists' fears that Burr would allow partisanship to influence the outcome of Chase's trial, his handling of the unprecedented proceedings was widely praised. Chase, who had certainly earned a reputation as an obnoxious justice, was acquitted on all counts. One Federalist senator later remarked, "I could almost forgive Burr for any less crime than the blood of Hamilton for [the] decision, dignity, firmness and impartiality with which he presides."[25] The Senate's acquittal of Justice Chase established a significant precedent in insulating judges against purely political assaults.

Following Chase's acquittal, Burr left Washington and the Senate, but not before delivering a farewell speech that witnesses said brought many listeners to tears. He concluded it with the words: "May the Almighty bless you and keep you in all that you do together here and separately in your own homes. I ask only that you not forget me, for I of a certainty, shall always remember, with respect and affection, the years I spent here."

On January 22, 1807, Thomas Jefferson delivered a message of his own to the Congress:

> Thomas Jefferson
> Special Message on the Burr Conspiracy
> To the Senate and House of Representatives of The United States.
> January 22, 1807.
>
> I received intimations that designs were in agitation in the western country, unlawful and unfriendly to the peace of the Union; and that the prime mover in these was Aaron Burr, heretofore distinguished by the favor of his country. The grounds of these intimations being inconclusive, the objects uncertain, and the fidelity of that

> country known to be firm, the only measure taken was to urge the informants to use their best endeavors to get further insight into the designs and proceedings of the suspected persons, and to communicate them to me.

When he left the Senate and Washington in March 1805, Burr was politically broken and nearly bankrupt. His law practice was finished. To satisfy creditors, he had been forced to sell off his estate, Richmond Hill, to an enterprising fur trader named John Jacob Astor—soon to become America's first millionaire—who broke the estate into smaller lots and leased them out, surmising correctly that Manhattan was soon going to grow. Out of power, out of friends, and seemingly out of ready cash, Burr began his "western adventure." Like so many other moments in his biography, the undertaking still confounds historians and generates controversy over his motives. But it was this adventure that brought him to a Richmond courtroom in the spring of 1807, and to the distinct possibility of hanging.

Like many Americans at the time, Aaron Burr saw what Jefferson's Louisiana Purchase of 1803 would mean for the future of America, and he turned his eyes westward. Like Jefferson, Burr anticipated that the transfer of wide tracts of North America from France to the United States, a transfer which left boundaries murky, would result in a war with Spain. At some point, he began to plan for what was then called a "filibuster," meaning an invasion by a private army to take over another nation's territory.*

*The modern meaning of "filibuster" is talking nonstop to block legislation in the Senate.

Although the national government considered such adventures a violation of the Neutrality Act of 1794, Burr's idea certainly fitted in with American ambitions to eliminate the Spanish presence from the continent and acquire the Spanish-controlled territory in what would eventually become Alabama, Mississippi, and Florida. Burr also envisioned liberating Mexico from Spanish rule. Whether he planned, as well, to sever the western states from the Union and establish a new nation is a far murkier question. "Burr's defenders insist that his aims were military"; wrote the historian Jean Edward Smith: "to provoke a war with Spain, to liberate Mexico, and ultimately free South America from Spanish rule. From this perspective, Burr was a patriot and his enterprise reflected the expansionist impulses of nineteenth-century America."[26]

Illegal, yes. But in a modern context, so was the plan hatched during the 1980s under the Reagan administration to support Nicaragua's contras against the leftist Sandinista government. When Congress made support of the contras illegal, officials in the CIA and in the Reagan administration, including Colonel Oliver North, looked for third-party funding for the contras, and this quest ultimately led to the Iran-contra scandal. Just as North was viewed by many Americans as a patriot who had broken a bad law in service to the country, Burr had many admirers who thought his aims were lofty.

To carry out his plan, Burr enlisted General James Wilkinson, who seemed in sympathy with it. The two had first met during the Quebec campaign in 1775. Despite what was at best a spotty military record, Wilkinson had been chosen by Jefferson to serve as the commander in chief of the U.S. Army at New Orleans and as governor of the Louisiana Territory. Neither Jefferson nor Burr knew that Wilkinson was already secretly on the payroll of Spain. Ignorant of Wilkinson's duplicity, Burr enlisted the general and others in his plan during a reconnaissance mission to the West in April 1805.

A man seemingly devoid of both competence and conscience, Wilkinson had, during the Revolution, survived several controversies, including his apparent participation in the unsuccessful "Conway cabal" that attempted to unseat George Washington. During the war, Wilkinson was given the role of clothier general of the army, but he was forced to resign that post amid charges of corruption in 1781. A hefty, hard-drinking man, Wilkinson was once described by the historian Frederick Jackson Turner as "the most consummate artist in treason that the nation has ever possessed."

After the war, Wilkinson moved to Kentucky and began to receive financial support from Spain. In 1787, he had actually sworn allegiance to the king of Spain, and he continued to receive Spanish funds for years. Even when he was returned to an army command during an Indian war in 1791, Wilkinson continued to provide Spain with intelligence about American plans for attacking New Orleans, then a Spanish possession. Astonishingly, in the "Quasi-War" of the late 1790s, he ranked third in the army, behind Washington and Alexander Hamilton. And it was Wilkinson who had the honor of accepting Louisiana from France in 1803; afterward, Jefferson appointed him governor of the Louisiana Territory.

When he was again removed from office for abuse of power, Wilkinson attempted to curry favor with Jefferson—and perhaps with his own Spanish paymasters—by revealing Burr's plans; but he did not explain to the president that he was himself a party to the "Burr conspiracy," as Jefferson would call it. Wilkinson's possession of a letter, the famous "Cipher Letter," prompted Thomas Jefferson to announce in his message to Congress that Aaron Burr was a traitor and his "guilt is placed beyond question." Jefferson vowed he would bring the full force of the federal government down on his former vice president. But Jefferson did not know that Wilkinson's damning letter was a copy that the duplicitous general had made, altering the text.

Watching with interest from a distance, John Adams thought that Jefferson had overreached and wrote to another signer of the Declaration, Benjamin Rush, that even if Burr's "guilt is as clear as the noonday sun, the first magistrate ought not to have pronounced it so before a jury had tried him." No admirer of Burr, Adams nonetheless also noted, "I never believed him to be a fool. Politicians have no more regard for the Truth than the Devil [and] I suspect that this Lying Spirit has been at work concerning Burr."[27]

BURR, IN THE meantime, was moving swiftly around the territories, while making contact with members of the British government, and seeking support for his ventures. And if he was truly on the verge of treason, he did little to cover his tracks. Instead he reached out to several prominent and powerful friends, such as his longtime acquaintance and Princeton classmate Senator Jonathan Dayton, and his son-in-law Joseph Alston.

He also made several overtures to a rising young power from the new state of Tennessee, forty-year-old Andrew Jackson. In May

1805, Burr had spent several days with Jackson and Jackson's wife Rachel at their home outside Nashville. Burr said that he intended to oust the hated Spanish from the Southwest, and Jackson was enthusiastic about the idea. Like many southerners and westerners, Jackson believed Spain was the enemy. The "dons," as he called them, contributed to two irksome and dangerous problems: Indians and fugitive slaves. Jackson and other southerners with their own ambitions of spreading American control looked upon Burr's plan very favorably.

Jackson believed that Burr was going to collect an expeditionary force in Kentucky and Tennessee and float down the Mississippi to New Orleans intending to start a revolution in Mexico, perhaps even with American naval support. Jackson's ardor for the plan began to cool, however, when he heard the rumors of Burr's real design: to create a new country with himself at the head. As the historian Andrew Burstein puts it: "Jackson was in charge of building boats for which Burr had paid in advance, at his Clover Bottom boatyard. Believing that Burr was not doing anything that Jefferson was unaware of—or else to cover himself—Jackson wrote the President, 'In the event of insult or aggression made on our Government from any quarter . . . , I take the liberty of rendering [Tennessee's volunteers'] service, that is, under my command.' "[28]

Jefferson issued an order for Burr's arrest, declaring him a traitor even before an indictment was issued. Burr turned himself in to the federal authorities. Defended in a Kentucky courtroom by a rising young attorney named Henry Clay, Burr was released after two grand juries separately found his actions legal. But Jefferson's warrant followed Burr to Mississippi, where still another grand jury refused to indict him in February 1807. Certain that the president and Wilkinson would not permit him to go free, Burr fled toward Spanish Florida. It was there that Nicholas Perkins recognized him late on the night of February 19, 1807.

Burr was brought to trial before the United States circuit court at Richmond, Virginia. Thomas Jefferson watched as his distant cousin and political antagonist, John Marshall, chief justice of the United States, presided over the trial. A staunch Federalist, Marshall had been appointed by John Adams just before Adams left office and in February 1803 had written the landmark decision in *Marbury v. Madison*, establishing the principle of judicial review, under which the Supreme Court can declare an act of Congress unconstitutional.

Playing a minor supporting role in Richmond, Andrew Jackson, who was never called as a witness, took every opportunity to publicly rail against Jefferson, call Wilkinson a liar, and contend that Burr was an innocent victim. The proceedings, which got under way in the spring of 1807, would involve weeks of pretrial motions and testimony before a grand jury stacked with Jefferson's allies, preceding the actual trial. It lasted until October 1807. But it still captivated political America. Convinced of Burr's guilt, eager to destroy the New Yorker who he thought had betrayed him in 1800, and also looking to damage Marshall, Thomas Jefferson gave the prosecutors blanket guarantees of pardons for anyone who testified against Burr.

Burr would lead his own defense, but his team of lawyers included Luther Martin, a Princeton classmate whom Burr had watched skillfully defend Samuel Chase in his impeachment trial. Martin had also been a delegate to the Constitutional Convention, where he had insisted on a resolution that prohibited further importation of slaves, or at least a tax on them if the trade was permitted. Martin said, "It was inconsistent with the principles of the Revolution and dishonorable to the American character to have such a feature in the Constitution."[29] Ultimately, Luther Martin was among those delegates at Philadelphia who refused to sign the Constitution.

Irascible and long-winded, Martin was nevertheless called the best trial lawyer in America. But he had a penchant for bourbon and was often obviously drunk in court, as he had been at the Constitutional Convention. In their history of that convention, *Decision in Philadelphia*, James and Christopher Collier offer a clear portrait of Martin's habits. "By the time of the Convention, he drank at times to the point of making a public spectacle of himself. It was, of course, a hard-drinking age, and the world was full of alcoholics. But gentlemen were expected to drink and stay reasonably sober, and Martin clearly went beyond the bounds of propriety."[30]

Although it took place in the early nineteenth century, the trial was a close equivalent of a modern "media circus." Richmond was filled with eager spectators, and there were reports that the town's population had doubled, to 10,000. And most of the journals of the day, almost all of which were aligned with one party or the other, were well represented. Among the "reporters" was a twenty-four-year-old New Yorker who wrote for a New York City newspaper edited by his brother and partly owned by Aaron Burr. An aspiring attorney, this writer had contributed to the *Morning Chronicle* under the pseudonym "Jonathan Oldstyle." But his real name, not yet well known in America, was Washington Irving.* He provided readers with a description of one of the most dramatic and eagerly anticipated moments in the courtroom: the arrival of General James Wilkinson, the chief government witness against Burr:

> *Wilkinson strutted into court, and took his stand in a parallel line with Burr on his right hand. Here he stood for a moment, swelling like a turkey-cock, and bracing himself up for the encounter of Burr's eye. The latter did not take any notice of him until the judge directed the clerk to swear General Wilkinson: at the mention of the name Burr turned his head, looked him full in the face with one of his piercing regards, swept his eye over his whole person from head to foot, as if to scan its dimensions and then coolly resumed his former position, and went on conversing with his counsel as tranquilly as ever. The whole look was over in an instant; but it was an admirable one. There was no appearance of study or constraint in it; no affection or disdain or defiance; a slight expression of contempt played over his countenance.*

The grand jury heard Wilkinson's testimony and saw the crucial piece of evidence: the letter Wilkinson said he had received from Burr. It was the only piece of physical evidence presented to

---

*In 1809, Irving published a satirical book, *A History of New-York from the Beginnings of the World to the End of the Dutch Dynasty*, under the pseudonym "Diedrich Knickerbocker." It was successful, but another ten years would elapse before he began to write the stories that made him famous, including the initial appearance of "Rip Van Winkle" in 1819.

the grand jury. This "Cipher Letter," allegedly written by Burr, proposed taking over territory that had been acquired under the Louisiana Purchase. But during the examination it was discovered that this key to the prosecution's case was in Wilkinson's own handwriting—a "copy," the pompous Wilkinson testified, because he had "lost" the original. The grand jury threw the letter out, and the news discredited Wilkinson and made him a laughingstock for the rest of the proceedings. Five days of cross-examination further demolished his reliability and his reputation. Thomas Jefferson had bet the house on Wilkinson and would lose badly. Other important prosecution witnesses failed to place Burr at the scene of the alleged conspiracy, an island in the Ohio River named for its owner Harman Blennerhassett, one of Burr's indicted coconspirators. And the credibility of the prosecution was further damaged because one of the men who was supposedly a witness to the conspiracy was never called to testify, even though he was sitting in the courtroom.

Overseeing these proceedings was a giant of American legal history. Born in rural Virginia in 1755, Justice John Marshall had, like Burr, served in the Revolution, including a winter at Valley Forge. Devoted to George Washington—he wrote a five-volume biography of the first president—Marshall was also a vigorous supporter of adopting the Constitution, and a staunch Federalist. He had been a "midnight appointee" to chief justice in 1801 by a lame duck President Adams, and he and President Jefferson became bitter rivals as Jefferson attempted to bend the judiciary to his will and Marshall resisted. In 1803, Marshall had irked Jefferson with the decision in *Marbury v. Madison*, establishing the principal of "judicial review"—the idea that courts could overturn and nullify the acts of other branches of government. In Burr's trial, Marshall would set more precedent. First, he allowed Burr to subpoena papers from Thomas Jefferson, establishing the fundamental concept that the presidential claim of executive privilege had limits and that the president was still answerable to the law.

On August 29, Chief Justice Marshall began work on his decision, which would really amount to a charge to the jury. Running to some 25,000 words, it was the longest decision he ever wrote in a storied and influential career. It took three hours to read in court, and then, after ruling on several points of law, Marshall turned the case over to the jury. One of the significant elements of his decision was a very narrow definition of treason, based on his reading of Article III of the Constitution. On Monday, September 1, the jury returned a verdict: "We of the jury say that Aaron Burr is not proved to be guilty under the indictment by any evidence submitted to us. We therefore find him not guilty."

With this verdict, the charges of treason against Burr's companions were also dropped. A second trial on a misdemeanor charge against Burr also resulted in a verdict of not guilty. Aaron Burr was a free man.

Summarizing the case and its importance, the legal historian Joseph Wheelan concluded, "Had a more pliable judge than Chief Justice John Marshall presided over Burr's treason trial, the Judiciary might have evolved into an instrument of repression, as it is in other nations. Rather than being interpreted as the Constitution's framers wished, the charge of treason might have become a means of silencing political opponents. The president and his official papers might today be immune from subpoena, and lie concealed behind a curtain of unassailable secrecy."[31]

Aftermath

In the wake of the sensational trial, Burr was left deep in debt. He was also under a constant threat of lynching by mobs of people who believed the worst of the press accounts. One of his codefendants, Harman Blennerhassett, was similarly in debt; his fortune, his home, and his property on the private island in the Ohio River where the alleged conspiracy had unfolded were all gone. Preparing to run for governor of South Carolina, Burr's son-in-law, Joseph Alston, provided Blennerhassett with funds that in retrospect seemed suspiciously like hush money. A broken man, Blennerhassett eventually moved to England and died there, bankrupt.

In June 1808, Burr also left for England, still hoping to get financial backing for another adventure, or "filibuster," against Spanish territories in America. He spent the next four years traveling around Europe, speaking, writing, and living on the edge of bankruptcy. But he was a shell of his former self. Once among the most powerful men in American politics, he no longer had any influence, and none of the European powers, including France under Napoléon, showed any interest in backing another of his schemes.

Finally he sailed back to the United States, disguised in a wig and whiskers and using an assumed name, Adolphus Arnot. After receiving assurances that he would not be arrested for debt, Burr returned to his former base of power, New York, at just about the time that the War of 1812 was declared. Despite his very checkered past, Burr opened a law office, and though very much a social outcast at first, he attracted clients who still valued his skills.

In June of that year, Burr learned that Theodosia's ten-year-old son, Aaron Burr Alston, had died in South Carolina. Crushed by the loss and also in poor health, Theodosia, decided to return to New York to visit with her father. Although travel was dangerous because England and American were at war, she boarded a fast ship, the *Patriot*, from Charleston, bound for New York. The trip should have taken five or six days. Burr went to the docks each day to await his beloved daughter's arrival. But Theodosia was never seen again. A bad storm had hit the coast and the ship was never heard from. There were rumors that pirates had captured the vessel. Burr continued to return to the docks for days until it was clear that there was no hope. He was heartbroken.

"When I realized the truth of her death, the world became a blank to me," Burr later said. "And life had then lost all value."

Burr lived on, often in debt. But he still proved a loyal friend when Luther Martin, his attorney at the trial, fell on hard times. Alcoholism had broken Martin, and Burr took his old friend and ally into his New York home, where Martin died in 1826, four days after both Thomas Jefferson and John Adams died on the fiftieth anniversary of the Declaration of Independence.

In 1830, Burr suffered the first of several strokes. Yet in 1833, at age seventy-seven, he married a former courtesan, Eliza Bowen Jumel, now fifty-eight and the widow of the wealthy wine merchant

Stephen Jumel. When she realized that her fortune was dwindling as a result of her new husband's land speculation, the couple separated, after only four months. Within a year, she had sued for a divorce; it was finalized on September 14, 1836, the day Burr died in the village of Port Richmond on Staten Island. He is buried in Princeton Cemetery near both his father and his grandfather.

In his last years, Burr told anyone who would listen that he had never harbored any treasonous ambitions. Before his death, having learned that America had recognized the independence of Texas from Mexico, he supposedly said, "There! You see? I was right. I was only thirty years too soon! What was treason in me thirty years ago is patriotism now!"[32]

In his landmark *History of the United States of America during the First Administration of Thomas Jefferson*, Henry Adams, son of one American president and grandson of another, aptly summed up the Burr case: "Never in the history of the United States did so powerful a combination of rival politicians unite to break down a single man as that which arrayed itself against Burr."

More than 200 years after Aaron Burr's trial, the story of his prosecution—or persecution?—remains timely and instructive. Recent history attests to the awesome power of an unchecked presidency in destroying a political enemy. The downfall of Richard Nixon in the aftermath of Watergate was largely a result of his concerted effort to destroy "enemies," an effort that involved not only the White House and the executive branch but the FBI and CIA as well. Nixon's "high crimes and misdemeanors" related to Watergate bordered on paranoia and certainly assaulted the Constitution. More recently, during the Iraq War, in the case of Ambassador Joseph Wilson and his wife, the CIA operative Valerie Plame, there was an effort to destroy a man's reputation and, in Plame's case, ruin a woman's career, if not threaten her life by exposing her clandestine work. After Ambassador Wilson had contradicted the Bush White House regarding Saddam Hussein's supposed attempt to acquire materials for an atomic bomb, Wilson was attacked through leaks to selected journalists and Plame's CIA cover was blown.

The government's ability to use the power of the prosecutor and grand jury remains one of the greatest threats to individual liberty. And the complicity of the press in these efforts is all too apparent. The means have changed with advances in technology; but the rules have not.

Burr's last measure of influence may have come from mentoring Martin Van Buren, a young New York attorney from the upstate hamlet of Kinderhook. They had first met in 1803, and in later years they worked together on several cases. Van Buren, who, like Burr, was short and also a "dandy" in his dress, was even rumored to be Burr's illegitimate son. He was not. But this speculation was another part of the continued attacks on Burr's reputation and legend that continued well in to the late nineteenth century. (Other writers would contend that Burr and his daughter Theodosia were incestuous, but no modern biographer takes this speculation seriously.) Nonetheless, Van Buren profited from Burr's tutelage and did become secretary of state and later, like his mentor, vice president during the presidency of Andrew Jackson, the future war hero who had come so close to joining "Burr's conspiracy."

II

Weatherford's War

1806 Andrew Jackson kills the attorney Charles Dickinson in a duel fought over a horse race on May 30.

1807 The first of three Embargo Acts is passed on December 22. In an attempt to keep America neutral during the Napoleonic Wars between France and England, these laws essentially ban all trade with foreign countries. They wreak havoc on the American economy.

1808 James Madison is elected the fourth president of the United States.

1809 The Shawnee chief Tecumseh and his brother, known as the Prophet, begin a campaign to create a Native American tribal confederacy to resist westward movement of American settlers. Tecumseh is allied with the British authorities in Canada.

1810 American settlers living in Spanish West Florida rebel against Spain, seizing the fort at Baton Rouge in September. In October, President Madison announces that this region is now part of the Territory of Orleans.

The census records the population of the United States at 7,239,881, a 36.4 percent increase since 1800; 1,378,110 are black, and all but 186,746 of those are slaves.

1811 The Battle of Tippecanoe. Governor William Harrison of the Indiana Territory defeats a surprise attack led by Tecumseh's brother, the Prophet.

1812 The Territory of Orleans is admitted to the Union as Louisiana, a slave state (the eighteenth state), in April.

The War of 1812 is declared on June 19.

President Madison is reelected in December.

1813 The Fort Mims massacre takes place in August.

1814 Andrew Jackson defeats the Creeks at the Battle of Horseshoe Bend in March, ending the Creek War.

Peace negotiations begin between the United States and Great Britain in Ghent. The Peace of Ghent will be signed on December 24, 1814.

1815 Andrew Jackson leads American forces to victory in the Battle of New Orleans, fought on January 8; it takes place after the peace accord has been signed.

1816 James Monroe is elected the fifth president of the United States.

But an evil day came upon us. Your forefathers crossed the great water and landed on this island. Their numbers were small. . . . We took pity on them, granted their requests, and they sat down among us. We gave them corn and meat; they gave us poison in return.

—Chief Red Jacket of the Seneca (1805)

Where today are the Pequot? Where are the Narragansett, the Mohican, the Pokanoket, and many other powerful tribes of our people? They have vanished before the avarice and the oppression of the White man, as snow before a summer.

—Tecumseh of the Shawnee (1810)

Jackson had his men cut off the noses of the fallen Indians to tally the body count. At the Treaty of Fort Jackson in August 1814, Jackson dictated peace terms and confiscated almost two-thirds of the Creek homeland, including lands belonging to the Lower Creeks who had assisted him in the war.

—Colin G. Calloway,
The Shawnees and the War for America

Mississippi Territory

August 30, 1813

On a sultry morning in late August, the stifling air—as thick as molasses and heavy as a wet wool blanket—slowed every movement and sapped most ambition. The weather was typical of the densely forested southern wilderness along the Alabama River. The trees in the nearby woods stood unmoved by any breeze. The heat shimmered off a nearby marsh.

This was Mississippi Territory, part of Thomas Jefferson's Louisiana Purchase, the extraordinary real estate deal that had doubled the size of America in 1803. Here, in what would eventually become the state of Alabama, white settlers had begun to arrive in growing numbers, a swelling tide of farmers eagerly looking for farmland and finding it in the dark, rich soil watered by the nearby Alabama and Tombigbee rivers. One day, this land would be part of King Cotton's domain, the black belt that would produce the American south's "white gold," enriching a few, enslaving millions of others.

But the future of this settlement held no such promise today. Those who had come to this spot, located some fifty miles north of the port of Mobile, arrived out of fear. In a clearing on about an acre of land at the edge of the dense southern forest and swamplands, they had thrown up a wooden palisade about seven feet tall, with loopholes through which defenders could fire their muskets. The wall was built around the farmhouse and land of a settler named Samuel Mims. Its name, Fort Mims, was deceptive. The simple barricade of sharpened posts, with its single large gate, could hardly be counted as a reliable, strong refuge. Its blockhouse, the central defensive feature, was still unfinished.

But the few hundred settlers who had sought refuge around Fort Mims still considered it a haven from the increasing threat of attack by hostile Creek Indians, who at that moment were at war, both among themselves and against Americans. The violence had grown out of a split within the Creek nation between two warring factions and eventually spilled over into attacks on the white American settlers who continued to stream into the territory, bringing thousands of slaves with them and stealing Creek land and challenging or destroying the Creeks' way of life. One faction of Creeks was willing to accommodate the whites and their ways. The other faction, known as Red Sticks for the color of its war clubs, wanted war.

Living in a scattering of cabins and other small buildings, the community was typical of America's southeastern outback in 1813. The frontier residents included a mix of white settlers; many Muskogee, or Creek, Indians who had completely adopted the farming and trading lifestyle of the whites; and a large number of other mixed-blood Creeks, descendants of decades of intermarriage and intermingling of Indians and Anglo-American settlers and traders.

In addition to the settlers and Indians, there were a significant number of black slaves, perhaps as many as 200, owned by both white settlers and Creek Indians, who had traditionally kept prisoners of tribal wars as slaves but had adopted the European preference for African slaves as well. The Creek were one of the so-called Civilized Tribes of the Southeast, many of whom had adjusted to the Anglo-American culture and economy, some of them even accepting baptism as Christians. There was also a small force of Louisiana militiamen assigned to provide security. In other words, Fort Mims was a curious melting pot of early-nineteenth-century frontier America.

Complicating the lives of the settlers, Creek Indians, and slaves at Fort Mims was the fact that the United States was at war. A little more than a year before, in June 1812, the nation had declared war on England. Although the news of that far-off conflict was slow in arriving and barely mattered to day-to-day life in Fort Mims, the war brought to the fretful settlers a constant stream of rumors: the Spanish and British were supposedly planning to ally themselves with hostile Creeks, arm 15,000 of the Indians with muskets, and set them loose on the American South, opening a new front in the war.

The idea that the British would stir up the Creeks was a nightmarish prospect to the American settlers. When a group of Creek warriors actually did travel to Pensacola, a fortified town in Spanish Florida, to receive some weapons promised by the Spanish, an American militia force attacked the Creek party in an otherwise inconsequential skirmish called the Battle of Burnt Corn Creek. Suddenly an all-out Indian war loomed much larger.

That is why the small detachment of Louisiana militiamen stood guard at Fort Mims. Their commander, Major David Beasley, may have thought all the talk of war was just that—talk. But when two slaves reported seeing some Indians nearby wearing war paint on August 29, 1813, Beasley dispatched some scouts. They found no sign of Indians, of trouble, or of any imminent attack. Beasley had the slaves whipped. His judgment would prove fatal.

At about noon on August 30, children danced in the open yard of the enclosed compound. A few soldiers played cards, and some of them reportedly got drunk. Lunch was set out for the gathering workers, who came to the fort at the sound of a military drum calling them to eat. Then, without warning, between 750 and 1,000 Creek warriors emerged from the tall grass and deep woods surrounding Fort Mims, howling their bloodcurdling war cries. With the suddenness and violent fury of a Gulf hurricane, they were on the fort. Carrying muskets, tomahawks, and the red-painted war clubs, they completely surprised the lone sentry posted at the open gate, who was distracted by the card game his fellow soldiers were playing nearby.

With the first shouts of "Indians," Daniel Beasley bolted toward the fort's gate, which was not only wide open, but also immovably jammed by large piles of drifting sand. Almost immediately, Beasley was shot in the stomach and gravely wounded, but he still tried to close the gate. It would not budge, and as Beasley fell, he desperately implored his soldiers to make a stand, even as the first wave of attackers quickly finished him off with their war clubs, now reddened by blood.

The Massacre at Fort Mims was on. The war party, led by the

mixed-blood renegade Peter McQueen, pushed into the stockade. The Red Stick warriors attacked head-on, having been assured by their tribal priests that they had magical protection from the weapons of whites. That assurance quickly proved wrong, as the historian Sean Michael O'Brien recounts: "Four Red Stick prophets, covered in feathers and black war paint, were first to enter the fort, with their warriors close behind. Confident that the bullets of the white men would simply pass harmlessly around their bodies, the prophets marched boldly through the gate. Three of them fell dead at once as the soldiers mustered their courage and opened fire."[1]

The militiamen inside the fort, along with the others in the settlement, vainly tried to fight back. Some took refuge in the main Mims house, at the center of the compound, which was partly surrounded by a low wall. As women loaded guns, the men fired at their attackers from the upper windows.

In a second building, the loom house known as "the bastion," another officer rallied his men to keep up a steady stream of musket fire that held some promise of turning back the Creek attackers. Women joined the fighting, including the stoic wife, now the widow, of Major Beasley: "Inside the bastion, Major Beasley's wife loaded weapons and urged the men not to surrender, and when a sick sergeant would not get up and fight, she became so enraged that she stabbed him with a bayonet."[2]

After three hours of intense hand-to-hand fighting in the sapping heat, the onslaught of the Red Stick warriors seemed to flag, and the assault died down. The Red Sticks learned that their principal chief, Far-Off Warrior, had been killed—and this news also seemed to take much of the starch out of their fight. But the other chiefs then chose William Weatherford, also known as Red Eagle, as their new chief. A mixed-blood Creek with a white father, William Weatherford had cast his lot with the Red Stick Creeks.

Red Eagle immediately called for a renewed assault on the stockade. Using axes, the Red Sticks, with an overwhelming number of fighters, broke through the compound's western wall, and a swarm of warriors poured into the enclosure. Red Eagle's warriors soon began to torch some of the houses. The Mims house, one of the best-defended buildings, was set on fire by burning arrows, and some of the women and children ran from it to the nearby loom house. As the frenzy of the attackers gained momentum, the grisly end of Fort Mims was approaching fast.

With their last refuge now completely ablaze, the remaining settlers were forced to flee the conflagration. As they ran for their lives, Red Stick warriors began the grim business of taking scalps. Finally, the air was filled with the sound of explosions as the powder magazine inside the Mims house exploded. For the settlers of Fort Mims, the apocalypse had come.

In the loom house, the last structure still being defended, Captain Dixon Bailey, half Creek himself, kept up a staunch resistance but realized that the odds were simply too great. With the building ablaze, he and the fort's assistant surgeon, Dr. Holmes, cut a hole in a wall. Accompanied by Bailey's ailing son, carried by a slave named Tom, and a black woman named Hester, Bailey and Holmes ran for the wood line. But Tom and the boy were quickly caught and brought down. Bailey saw the attacking Indians smash both of their heads with war clubs. Continuing to run from the compound, Bailey, Dr. Holmes, and Hester made it to a nearby swamp. Wounded and bleeding, Bailey died there. The doctor found a hiding place in the marsh, while Hester walked the twelve miles to Fort Stoddert and reported what had happened at Fort Mims.[3]

After Bailey's doomed breakout, the last of the resistance inside the fort quickly collapsed. The Red Sticks' advantage in numbers and arms was too great. The warriors finally battered down the entry to the bastion and fell on the last of the settlers. Dragged into the open yard amid the burning buildings, the exploding ordnance, and the screams of the dying, most of the surviving whites and mixed bloods alike were slaughtered without mercy.

In a grim summary of the fighting and its aftermath, Sean Michael O'Brien records, "Small children were taken by the feet and dashed against the walls of the fort, splattering the walls with brains and blood. Women were stripped, scalped and mutilated. Pregnant women were ripped open and their unborn babies killed before their eyes."[4]

According to later accounts, William Weatherford—Red Eagle—attempted in vain to stop the killings of the women and children. But the bloodlust was too high. Later, Weatherford would say of the deadly assault on Fort Mims: "My warriors were like famished wolves and the first taste of blood made their appetites insatiable."[5]

Perhaps forty defenders escaped to tell the tale of the Fort Mims Massacre. Some Red Stick warriors recognized familiar faces among the survivors and took them captive instead of killing them. The slaves who survived were marched away, simply to become slaves of the Red Stick Creeks. More than 250 of the inhabitants of Fort Mims were killed or taken captive.

When a relief column from Fort Stoddert finally arrived days later, they found the grim evidence of the horrific massacre. The soldiers buried at least 247 bodies, many of them scalped, dismembered, and charred from the fires. The attack on Fort Mims—the worst frontier massacre in American history—and its aftermath would have extraordinary consequences for the country.

WHAT HAPPENED AT Fort Mims was a result of more than a simple act of Native American rage at the white incursions into their lands or even of a tribal power struggle spiraling out of control. It was a complex and volatile mixture of competing territorial ambitions, international intrigues, greed, power plays, the vain attempt at survival of a culture and people, and warring civilizations. But mostly it remains a quintessential American story: a struggle over real estate—having it, wanting it, and fighting for it, sometimes to the death. The terrible violence that swept over Fort Mims was neither unusual for its day nor one-sided.

The story of Fort Mims and the Creek nation would be largely overlooked when the history of this period was written. But this horrific assault was as dramatic and devastating an attack as Pearl Harbor or 9/11 was for later generations of Americans. Even though the nation was at war with England, shocked and angry Americans demanded vengeance; it would come swiftly, decisively, and with a fury that nearly matched the violent atrocities of that day in Fort Mims.

The bloody fighting and conflict that followed—a small "war within a war" fought during the War of 1812—were relatively brief, a matter of months. In reality, there was a series of localized battles between American militia forces and portions of the Creek nation. To understand the Fort Mims Massacre and its place in American history, it is probably best to understand the two men who would oppose each other in the Creek War. One of those men might have been a savior and liberator to his people—had he prevailed. He was charismatic, daring, and ruthless in his own way—as much a folk hero to his Creek followers as William Wallace, depicted in *Braveheart*, might seem to Scots and to modern moviegoers. William Weatherford, also called Red Eagle, wanted what William Wallace, or George Washington, for that matter, wanted: freedom and his peoples' land. But he lost the war and was largely forgotten by the history books.

His opposite was equally daring, resolute, and certainly ruthless. He became America's seventh president. Andrew Jackson's exploits at the Battle of New Orleans, the decisive American victory over some of England's best troops in January 1815—after the War of 1812 was officially over—embellished Jackson's reputation as the greatest American general since George Washington. But that is not where the legend of Jackson began. It really started when he pursued the Indians who had attacked Fort Mims, promising as he set out, "Long shall the Creeks remember Fort Mims in bitterness and in tears."

It was not an empty threat.

The man who would track down William Weatherford was not given to words that were not backed by action. Since childhood, he had seen the violence of the world and often responded in kind. "Commanding, shrewd, intuitive, yet not especially articulate, alternately bad-tempered and well-mannered, Jackson embodied the nation's birth and youth," wrote Jon Meacham in *American Lion*. "Jackson was fond of well-cut clothes, racehorses, dueling, newspapers, gambling, whiskey, coffee, a pipe, a pretty woman, children and good company. . . . Depending on the moment, he could succumb to the impulses of a warlike temperament or draw on his reserves of unaffected warmth."[6]

Born in 1767, in South Carolina, Andrew Jackson had survived a brutally harsh and violent early life, coming of age in the unforgiving world of the colonial frontier and witnessing some of the most savage warfare of the American Revolution. His Scots-Irish father, also Andrew, arrived in America in 1765 with his wife Elizabeth and their two sons: Hugh, two years old when they arrived; and Robert, aged six months. While working the fields on a hardscrabble farm, the father dropped dead in February 1767, leaving behind a pregnant wife and the two young boys. Elizabeth Jackson moved in with relatives, and Andrew was born a month later, on March 15, 1767. He spent his early years in the home of relatives, the Crawfords.

Obviously bright, but with a boyhood reputation as a bit of a hell-raiser, Andrew Jackson balked at school. His mother's hope that he might go into the ministry was far from realistic. As Jackson's biographer Robert Remini once noted, "For one thing he swore a blue streak, fine, lovely, bloodcurdling oaths that could frighten people half to death. Also, he was wild and reckless. . . . There was an air of uneasy restlessness about him, an exuberance that found outlets in outrageous tricks and games. Every now and then, he showed an ugly side that labeled him a bully. Although not a coward, he would purposely terrorize people if angered or if it suited his needs."[7]

When the Revolution came to South Carolina, Andrew Jackson saw war and its depredations at their worst. In 1779, his brother Hugh, then sixteen years old, was killed in battle at Stono Ferry. The following year, British forces under the command of Lieutenant Colonel Banastre Tarleton—a notorious officer known as "Bloody Ban" or "the Butcher"—descended on Charleston, South Carolina. They laid waste to the town, by many accounts inflicting a vicious massacre that left more than 100 Americans dead, many after they had tried to surrender.

Then, in April 1781, Andrew and his brother Robert were taken prisoner. There is a story—perhaps apocryphal—that a British officer ordered the Jackson boys to clean his boots. Both defiantly refused, supposedly asserting that as prisoners of war they weren't required to do this. Jackson later claimed that the officer hit him and Robert with a sword for their defiance. They were then taken to a British prison camp in Camden, where their mother eventually won their release by pleading with the British. But on the trip home, Robert died; the head wound he had received from the officer's blow had become infected.

Not long after that, Jackson's mother went off to nurse some American prisoners who were being held on a British ship in Charleston harbor. Andrew Jackson never saw her again. Elizabeth Jackson fell ill with cholera and died in 1781, the year that Cornwallis surrendered at Yorktown and the war came to an end. At age fourteen, Andrew Jackson was orphaned. He had lost two brothers and his mother to the war; he had never known his father. His nightmarish childhood clearly hardened him. But as Robert Remini remarks, "No one has ever seriously questioned Jackson's courage or his sense of duty. Those who knew his family background understood a little about where and how he had obtained them."[8]

These war experiences also made Jackson intensely distrust those forces he saw as enemies of America. In his experience, they were the British, the Spanish, and the Indians. For the rest of his life, he hated all three with a passion that often drove his questionable actions.

LIKE MANY OTHER Native Americans of his day, William Weatherford saw the American government, and most whites, as the enemy. Long before the American Indian wars took place on the Great Plains in the late nineteenth century and became a staple of dime novels and Hollywood, long before the "pony soldiers" battled the Sioux, Weatherford and other Indians in the eastern and southern states had fought furiously against the encroachment of white American settlers. They were battling history and modernity.

Born around 1781, William Weatherford was the son of Charles Weatherford, a Scottish merchant and horse breeder who ran a trading post and racetrack. William's mother was a Creek of noble lineage. Such ancestry was not at all unusual. A roster of the names of a number of significant Creek tribal leaders of this generation reads like a gathering of the Highlands clans: they included Peter McQueen; William McIntosh, Weatherford's cousin; and Alexander McGillivray, Weatherford's uncle, who had earlier signed a peace treaty with George Washington's secretary of war, Henry Knox.

As the nineteenth century opened, many Creeks lived in settled towns and villages in Georgia and the future Alabama. Part of the Muskogee (or Muskogean) speaking groups in the American Southeast that included the Chickasaw, Choctaw, and Alabama nations, they had been forced from their traditional territories with the arrival of the Spanish and later the English in the American Southeast. But they remained a force in the region, and in 1730 a Creek trade delegation had traveled to London to negotiate agreements with the British government. Demolishing the simplistic image of the fur trade as a lopsided deal in which "civilized" Europeans exchanged trinkets for furs and skins with unsophisticated natives, the historian Daniel K. Richter explains, "A series of fatal bovine epidemics struck continental Europe, creating a huge market among leather workers for North American deerskins to replace now-scarce cattle hides. The Creeks—controlling territories that, largely as a result of their own previous slave-raiding expeditions, were devoid of humans but thronging with white-tailed deer—were ideally placed to profit from that demand."[9]

Having allied themselves with the British during the Revolution, the Creeks felt betrayed when they learned that the English had simply transferred ownership of their territory to the new American nation under the terms of the peace treaty that ended the Revolutionary War in 1783.

Under an official United States federal policy of "civilization," many Creeks and members of the other nations of the Southeast accepted the new way. As Richter explains, the U.S. government, "sought to teach Indian peoples to abandon their traditional gendered economy of male hunting, female agriculture and communal landholding in favor of male plow agriculture and animal husbandry, female domesticity, and especially, private property. This shift toward a Euro-American way of life, the theory went, would allow Indians to prosper on a much smaller land base, opening up the vast remainder to White yeoman farmers. Of course, it also envisioned the end of Indian culture and Indian political autonomy."[10]

The shift created policies that soon had Indians heavily in debt, which forced the tribes to cede enormous tracts of their lands to the federal government. In the midst of these increasingly corrupt and often underhanded dealings, aggressive white settlers were also raising tensions by swarming into the contested territories. Tribal lands, once held communally, were swallowed at an alarming rate, and although the Creeks attempted to accommodate white ways, their existence as a people was under assault. Also, to the typical white settler, the Indians were for the most part "savages," with no rights to land to which they had no deeds.

This was the crisis facing William Weatherford's Creek people as the War of 1812 approached. A turning point in Weatherford's life and in the future of the Creek nation came with the arrival in Creek territory of Tecumseh, a charismatic Shawnee leader who was trying to unite the disparate Native American tribes against the American government. Tecumseh had been born around 1768 in what is now Dayton, Ohio. Part Creek on his mother's side, he grew up at war with Americans. As a teenager, he joined the British during the Revolution. After the war, he fought on the losing side in several of the notable battles that secured American control over the future states of Ohio and Indiana.

While living in Ohio, Tecumseh apparently fell in love with

a white woman: Rebecca Galloway, the daughter of a farmer. She taught him history, Shakespeare, and the Bible.[11] But she wanted him to abandon his traditional ways, and Tecumseh refused. Around 1805, Tecumseh and his brother, Tenskwatawa—usually called "the Prophet"—became part of a militant Native American religious revival in which all tribes were urged to reject white ways and stop ceding lands to the United States. Just as the Great Awakening had transformed colonial American society and politics in the 1740s, this Native American "awakening" was succeeding in creating a united front against the inexorable westward movement of Americans into Indian lands in the Midwest.

Tall, impressive in his demeanor, and a natural orator, Tecumseh set out on a mission to create a confederation of Indian nations that could roll back the American advance and restore native traditions. Between 1809 and 1811, he began to preach "red unity" as a means of survival. Tecumseh provided political leadership, and his brother, the Prophet, gave their mission the feel of a religious crusade. He appealed mainly to the "Young Turks," angry young Indian men who did not want to be forced out of their land by the white settlers.

"Tecumseh's movement was a coalition of warriors, not of tribes," writes the Dartmouth historian Colin Calloway. "Warriors from far and wide cast aside their venal chiefs and gravitated to Tecumseh and his vision of a still-strong Indian nation that would stand up to American aggression."[12]

In 1811, Tecumseh took that vision to the Southeast, holding a series of tribal councils with leaders of the Civilized Nations, including the Creeks. His appeal merely divided the Creek leadership between two rival camps: those who wanted to join Tecumseh—

and some actually did ride off with the Shawnee general—and the "accommodationists," who thought that the prospect of war with America was suicidal. William Weatherford was initially caught between the two, but he would eventually be brought to the side of the warriors. This fundamental fissure within Creek ranks eventually contributed to a Creek tribal civil war that, fatefully, led to the attack on Fort Mims, and after that, the wider Creek War.

While Tecumseh was still on his recruiting mission through the Southeast, however, events spun beyond his control. His brother, the Prophet, who lacked Tecumseh's military genius and tactical skills, was contending with an American army led by the governor of the Indiana Territory, William Henry Harrison. The son of Benjamin Harrison, a wealthy Virginia planter who had signed the Declaration, William Harrison would become the country's ninth president.*

In November 1811, Harrison's 1,000-man force arrived at the Prophet's town, known as Tippecanoe. That night, the Prophet ordered an attack on Harrison's camp. Although Harrison's losses were far greater than those of the Indians, the Indian forces retreated and Harrison was able to claim a victory at Tippecanoe, burning the Prophet's deserted village and securing his own status as a bona fide war hero and Indian fighter. (And providing his memorable 1840 presidential campaign slogan: "Tippecanoe and Tyler too.")

This setback weakened the Prophet's standing and credibility because his prediction that American bullets would not harm his

*The first president to die in office, Harrison was also the grandfather of Benjamin Harrison, the twenty-third President.

warriors was clearly wrong. Following this defeat, Tecumseh decided to ally his forces with the British once more. When the War of 1812 began, Tecumseh joined a British major general, Isaac Brock, in forcing the surrender of a fort and town, both known as Detroit, in what was then Michigan Territory. But in October 1813, Tecumseh was killed at the Battle of Thames in Ontario. According to popular reports at the time, Tecumseh's body was dismembered by American troops, but no evidence of his remains was ever discovered.

Tecumseh's death was also a fatal blow to the realization of a great American Indian confederacy that might have delayed or staved off the press of white settlers flowing west. But Tecumseh's impact outlived him. "Although Indian hopes for holding the Northwest had died with Tecumseh, he had spread his word in the South more effectively than he knew," the historians Robert Utley and Wilcomb Washburn wrote. "Even while he was making his last stand on the Thames, Indians a thousand miles away who had been inspired by his rhetoric were beginning a struggle that would last nearly thirty years."[13]

Following Tecumseh's lead and his vision, the Red Stick Creeks also turned to the British as allies. With war under way against the upstart Americans, British agents in the Southeast openly recruited Indians and fugitive African Americans. By July 1813, the British and Spanish in Florida were providing weapons and powder to the Creeks and other tribes. The worst fears of the Americans living in the southern territories bordering Florida had come true. That was what sent a force of Alabama militiamen against the Red Sticks at Burnt Corn Creek. When the Creeks won that minor skirmish, it opened the floodgates for an open war against white settlers.

And that war would become part of the larger war fought between the British and Americans, a conflict that the historian Walter Borneman once described as "a silly little war—fought between creaking sailing ships and inexperienced armies led by bumbling generals. . . . In the retrospect of two centuries of American history, however, the War of 1812 stands out as the coming of age of a nation."[14]

WAR AS A coming-out party had never been on Jefferson's presidential agenda. As a matter of national policy and personal philosophy, he aimed to avoid "foreign entanglements." After the successful acquisition of the Louisiana Territory from France in 1803, Jefferson was reelected in 1804 in a landslide victory. The opposition Federalists of the defeated Adams and the late Alexander Hamilton were nearly extinct as a viable national opposition party.

For Jefferson, the only unpleasantness hanging in the air during these heady days came from the charge that, as the master of Monticello, he was involved in a sexual relationship with a slave, Sally Hemings. The localized whispers of the relationship became public in 1802, when James Callender, writing in the *Richmond Recorder*, first accused Jefferson of having had several children with Sally Hemings, who was also the half sister of Jefferson's late wife, Martha. Callender had revealed Alexander Hamilton's liaison with Maria Reynolds in 1797, and was thus responsible for Hamilton's downfall. A practitioner of the down and dirty mudslinging politics of his time, Callender had worked for Jefferson's campaign and produced acidic anti-Federalist propaganda during the elections of 1796 and 1800. But when he had requested a patronage job in the new Jefferson administration, the scandalmonger was rebuffed. For that and other reasons, Callender began his very public campaign in the press about Sally Hemings.[15] The opposition Federalist newspapers tried to make campaign hay of this story, but the budding scandal made no difference to the outcome in 1804.

Still, Jefferson might have been contemplating Callender in his second inaugural:

> *During this course of administration, and in order to disturb it, the artillery of the press has been leveled against us, charged with whatsoever its licentiousness could devise or dare. These abuses of an institution so important to freedom and science, are deeply to be regretted, inasmuch as they tend to lessen its usefulness, and to sap its safety; they might, indeed, have been corrected by the wholesome punishments reserved and provided by the laws of the several States against falsehood and defamation; but public duties more urgent press on the time of public servants, and the offenders have therefore been left to find their punishment in the public indignation.*

With his overwhelming victory, Jefferson thought that the two-party system was finished and that only peace and prosperity lay ahead: "The new century opened itself by committing us on a boisterous ocean. But this is now subsidiary, peace is smoothing our paths at home and abroad, and if we are not wanting in the practice of justice and moderation, our tranquility and property may be preserved."[16]

But it was not to be. In the early nineteenth century, America was still a small, untested, ill-defended nation—practically a defenseless and powerless pawn in the affairs of Europe's great powers. America's weaknesses and vulnerabilities were on display as Napoléon—who had sold off Louisiana at fire sale prices to raise cash for his wars—turned much of the world into a battleground between 1803 and 1815. Jefferson desperately wanted to keep the United States, with its paltry army, skeleton navy, and starving treasury, out of these European wars by remaining neutral. Initially, many Americans had actually profited from the Napoleonic wars, as the French and English eagerly sought American goods and ships.

But declarations of neutrality by the United States did not prevent its merchant ships from being stopped on the high seas, often by British vessels, whose officers could take any British subject and "impress" him into the service of the Royal Navy. On board ships at sea, such polite legalisms as "naturalized citizen" were meaningless concepts. As American sailors were being seized on the high seas along with British subjects, there was a rising anger at the British.

In 1807, Jefferson got Congress to agree to the first of three Embargo Acts, which prohibited all exports into America. Jefferson hoped to use economic retaliation against the British impressment policy and to show the two warring European empires how important their trade with America was. Jefferson's plan backfired badly; the Embargo Acts proved to be among the most unpopular, unsuccessful, and costly laws in U.S. history. Instead of teaching the British and French a lesson, the embargo merely hurt American merchants and devastated the American shipping industry. In particular, New England, where the old Federalist Party retained its last vestiges of power, was hard hit. Smuggling soon became rampant throughout America. Just as Prohibition in the twentieth century would lead to an explosion of organized crime, the Embargo Act "incentivized" lawbreaking.

By the end of his second term, Jefferson recognized the disaster, and sought to fix it by replacing the Embargo Acts with the Non-Intercourse Act, which lifted all embargoes on American shipping except those destined for British or French ports. It, too, failed miserably to halt the continued kidnapping of Americans at sea, and the economic hardship for American merchants and shippers continued to worsen.

When James Madison succeeded Jefferson in March 1809, war with England and perhaps with France seemed likely, if not inevitable. The coming war was set against the other constant threat—Indian uprisings. Spain's involvement in stirring up the Creeks was precisely the reason why so many southerners—Andrew Jackson vocally among them—hated the nation that still controlled Florida and why they had been so willing to embrace Aaron Burr's cause a few years earlier.

Thomas Jefferson may have temporarily eased matters when he bought Louisiana from France in 1803. But there were vast portions of mostly unexplored North American lands still up for grabs. The competition among the United States, England, Spain, and France—as well as the Native nations who still occupied large pieces of that territory—was fierce, and sufficient cause for war. In the U.S. Congress, a vocal group advocating war with England over impressments and the British support of Indians became known as "War Hawks." This group was led most conspicuously by Henry Clay of Kentucky and John C. Calhoun of South Carolina.

Following the official declaration of war in June 1812, much of the fighting had taken place around the Great Lakes and upstate New York, as the United States focused its military attention on British armies moving out of Canada. The only action near Fort Mims was the relatively uneventful and bloodless capture of the fort at Mobile, then a Spanish outpost on the Gulf of Mexico.

The first city built in the Louisiana Territory, Mobile had been founded by the French in 1710 and was the capital of the Louisiana Territory before being surpassed in significance by New Orleans. Mobile had changed hands several times over the years but was under Spanish control in April 1813. Assaulted by a modest force of only sixty men, the Spanish fort there surrendered without a shot being fired. The conquering American forces were led by none other than General James Wilkinson—still in the employ of Spain as "Agent 13."

The unlikely survivor of a disastrous performace at Aaron Burr's trial, Wilkinson had already dodged one court-martial and a series of congressional investigations into his behavior and his highly questionable dealings in Louisiana. But with more lives than the proverbial cat—and an unmatched ability to cover his traitorous tracks as a paid agent of Spain—Wilkinson was commissioned a major general as the war with England opened. Despite his spotty record, he was one of the few Americans with actual military experience. Later in the war, he was brought before a military court yet again after his defeat at the hands of a much smaller British force near the Canadian border. Wilkinson was never given another command, and eventually he returned to his plantation near New Orleans. He died in Mexico City awaiting a land grant in Texas in 1825. His one significant contribution to the American cause was keeping Mobile and its harbor out of British hands.

But after the Fort Mims Massacre, that American victory was small comfort to the people of the Alabama Territory. Fear bordering on panic quickly spread throughout the Southeast. From Louisiana through Georgia and Tennessee, emotions ran high. In Nashville, Tennessee, the state legislature called for 3,500 volunteers to respond to the massacre. Among the frontiersmen eager for vengeance was David Crockett, who wrote to his wife, "My countrymen had been murdered, and I knew that the next thing would be, that the Indians would be scalping the women and children all about there if we didn't put a stop to it. . . . The truth is my dander was up, and nothing but war could put a stop to it."[17]

At the Hermitage, his plantation near Nashville, Tennessee, Andrew Jackson, now wealthy and influential, clearly shared Crockett's views. In a letter about an alleged Creek attack on white settlers, sent to Thomas Jefferson in 1808, Jackson wrote, "These horrid scenes bring fresh to our recollection the influence, during our revolutionary war, that raised the scalping knife and tomahawk, against our defenseless women and children. The blood of our innocent citizens must not flow with impunity—justice forbids it and the present relative situation of our country with foreign nations require[s] speedy redress, and a final check on these hostile murdering Creeks."[18]

But when word of the Fort Mims Massacre reached Nashville in 1813, Andrew Jackson seemed to be in no shape to lead the Tennessean troops who would respond to the desperate pleas for help. At the moment, he was recovering from a serious wound—the result of a gunshot he had received in a street fight with two brothers, Thomas and Jesse Benton. Thomas Hart Benton was actually Jackson's protégé and aide. But in frontier America, family ties were thicker than political loyalty and Thomas Benton had sided with his brother, Jesse.

The fight had begun over an almost trivial argument that led to a duel between Jesse Benton and William Carroll, another friend and subordinate of Jackson's. The duel left both men alive—but Jesse Benton had a pistol ball in his buttock. The wound was painful and embarrassing. On September 4, 1813, just a few days after the Fort Mims Massacre, Jackson encountered Thomas Hart Benton—who would go on to become a powerful Senator from Missouri—and his brother, Jesse Benton. Standing by the City Hotel in Nashville, the men soon had words over the duel, and Andrew Jackson drew his gun. As some of Jackson's friends and family members passed by the scene, the situation escalated into a pitched gunfight and knife fight. Sometime during the course of the melee, Jesse Benton shot Jackson.

No stranger to duels and violence, Andrew Jackson had already killed one man, Charles Henry Dickinson, in a duel in 1806. Dickinson and Jackson had fought, it is generally agreed, over a horserace, and their argument led to words about Jackson's controversial marriage. In 1791, Jackson had married Rachel Donelson Robards, who believed that she had been divorced from her first husband. But she and Jackson had to be remarried in 1793 after it became known that her divorce papers were never finalized. The insults and scandal about their supposed "adultery" would dog Jackson throughout his political career, and the assaults on Rachel's honor infuriated him.

Tennessee had banned duels, so the two men went to neighboring Kentucky to settle their score. Dickinson fired first and hit Jackson just beside his heart. Jackson's pistol then misfired, so he calmly reloaded and shot Dickinson in the abdomen. Dickinson died hours later, in agony. Dickinson's bullet remained in Jackson's chest for years, causing abscesses that plagued him for the rest of his life.

After the gunfight with the Benton brothers, Jackson lost a great deal of blood. The doctors who treated him recommended amputating his arm below the shoulder wound, but Jackson refused. Weeks later, he was still recuperating and very weak when panicked settlers began spreading the horrific news of the Fort Mims Massacre. Gaunt and in excruciating pain, the forty-six-year-old Jackson was nonetheless ready for action. Despite his limited experience of field command, Jackson had inspired a devoted following among the Tennessee militiamen, who called him Old Hickory for his toughness. Marching toward Creek lands with a force of 1,000 infantrymen and some allied Creek and Choctaw Indians, Jackson pursued a ruthless "scorched earth" policy, attempting to destroy the Creek food stores and uprooting their villages. This was not war against an enemy army but an assault on the civilian population that supported the enemy.

Jackson's campaign against the Red Stick Creeks was troubled from the outset. Occasionally doubling over in pain as he rode, Jackson was also bedeviled constantly by the failure of the Tennessee state government to support his troops. Supplies, especially food, for the men and horses were a persistent problem. Jackson's soldiers were reduced to meager rations, and morale plummeted. Eager for vengeance on the Creeks and spoiling for a quick fight, many of Jackson's militiamen were untrained and proved poor soldiers. Giddy and overconfident when they set out, they eventually turned mutinous as the campaign stretched into weeks of scanty rations, difficult marches through tough wilderness terrain, and long stretches of inaction.

The situation began to change in November 1813, when Jackson was finally able to bring his troops into combat against a Red Stick force at the village of Talladega. Jackson had planned well, and the Red Sticks were lured into an open field, where Jackson's troops made short work of them. Davy Crockett later wrote, "We shot them like dogs."

In the aftermath of that first battle, a Creek infant was discovered near the body of his dead mother. Some of the troops discussed returning the child to other Creek women. Others suggested killing him. Jackson rode up and intervened. He had the infant taken to his tent, and later the boy was sent back to Nashville, where he was raised in the Jackson home as Jackson's adoptive son. Perhaps it was the recollection of having been a war orphan himself that softened Jackson on this battlefield. But the picture of the hardened commander, on a battlefield, stepping in to save a war orphan adds to the complex picture of his character. On the other hand, the boy was far from the last orphan created by Jackson's military and political decisions.

Jackson's fiery temperament and unforgiving approach to war were on display in this battle, the first action in what would be called the Creek War. The assault on the Creeks had been carried out with brutal efficiency. But there was more to come as the campaign dragged into winter. When the Creek warriors melted deeper into the southern wilderness, Jackson became more and more frustrated by his inability to meet and engage the enemy. Pushing farther into Red Stick territory, Jackson unleashed what has been described as "a widespread assault of savagery against the Indians, killing, burning villages, and plundering food supplies."[19]

The fact that runaway slaves, including some of those who had been taken from Fort Mims, were now allying themselves with the Red Stick forces added to Jackson's resolve. A slaveholder himself, Jackson would remain a staunch defender of the "peculiar institution," and he certainly did not want to encourage any runaways. These blacks had taken up the Red Stick cause as a means to their own freedom. The growing alliance between black men and red men was going to be one of the complicating factors as the Indian wars in the Southeast played out. Not only was Jackson sending a clear message to Indians of the dangers of taking up arms against America; he was also letting American slaves know that escape and rebellion were not options.

Jackson's attitudes toward Indians and slaves stand high among the reasons that some contemporary historians—and many Native Americans—take a dim view of the general that the Indians called "Sharp Knife." The historian David S. Reynolds has described the seventh president as "a potent killing machine," adding: "His shortcomings reflected his era, as did those of other great leaders from Jefferson to Lincoln. But understanding Jackson, perhaps more than most leading Americans of his time, requires an ability to resist either vilification or veneration to see the man whole—his failings as well as his successes."[20]

Despite his victories in the field, Jackson had yet to deliver the fatal stroke to the main Red Stick forces. As time went by, his plans were also threatened by growing troubles within his own ranks. Grumbling over the poor rations and endless marches increased daily. A few men had deserted, and with supplies short as winter approached and enlistments for some of these militiamen were coming to an end, Jackson was ultimately forced to confront a mutiny. With the support of a few loyal officers and men, Jackson had faced down a large group of militiamen who were about to desert. Jackson's unwavering, ramrod-stiff resolve as he stood in front of a large contingent of armed mutineers served him well. His courage and refusal to make concessions added to Old Hickory's legendary fearless determination. Just as he had shown the doctors his force of will when he refused amputation, his unwavering determination thwarted an armed revolt that might have ended the campaign, and with it his own career.

Sick and seriously weakened by March 1814, Jackson finally received some long-promised reinforcements and supplies. This timely relief column permitted him to maneuver his army for what he thought might be a climactic end to the war. At Horseshoe Bend, where the Tallapoosa River in what is now central Alabama curves in a U-shape, Jackson discovered Red Eagle's main force. With some 3,000 men, and more Indian allies from the Cherokee and Allied Creek nations, Jackson confronted the Red Sticks, about 900 warriors and 300 women and children, in their village by the river. On the north bank of the river, the Creek camp was well protected by a strong breastwork of logs and earth. The Red Sticks also had canoes by the river so as to escape if necessary.

On March 27, 1814, Jackson commenced the assault that would end the Creek War. Sending one force to the river's south side, or the rear of the Indian encampment to cut off retreat, Jackson opened his attack with an artillery barrage that did little to break through the village's well-constructed defenses. Then he ordered an infantry assault on the defensive works. At the same time, the troops he had sent to the rear of the village had begun to cross the river and had taken the Red Stick canoes—here again, to cut off retreat. As they moved in on the village from two sides, Jackson's forces held the Red Sticks in a deadly vise.

There was desperate hand-to-hand fighting as the Red Stick warriors fought to defend their women and children. Among the men who mounted the breastworks was a young officer named Sam Houston, a twenty-one-year-old Tennessean who had traded farming, and clerking in a general store, for the army and who would now take an arrow in his leg. Known to the Cherokee as Raven, he had lived with them and spoke their language. As the young officer had the arrow pulled from his leg, Jackson rode by and ordered him from the field. But Houston later returned to the battlefield of his own accord and attempted to lead another charge.

Against overwhelming odds and American artillery, the Red Stick defenses were inadequate; they were overrun, and the battle turned into a rout and finally a bloodbath. Warriors ran to the river to escape and were easily picked off by Jackson's soldiers until the river ran red.

Simple victory was not enough for the Americans, many of whom wanted to avenge the Fort Mims Massacre. "Many white soldiers mutilated the dead Muscogees, cutting off long bands of skin from the bodies to make belts and bridle reins for their horses. . . . Seeking a body count from the previous day's slaughter, Jackson's soldiers moved among the dead and sliced off the nose of each

Red Stick corpse to keep an accurate number. Old Hickory's officers counted 557 enemy dead—and estimated 250 to 300 more warriors killed in the river, raising the total to close to 850. Jackson's victory was complete and devastating."[21]

It was perhaps the bloodiest defeat of Native Americans in the long and tragic history since Europeans had first arrived. And America had a new war hero.

But William Weatherford was not among the dead. A few weeks later, a lone rider entered the newly constructed Fort Jackson. He approached Jackson's tent and announced, "I am Bill Weatherford."

At first enraged, Jackson demanded, "How dare you . . ."

But something in Weatherford's pluck and demeanor won Jackson's admiration. In an account that has rightfully raised the suspicion of some historians, Jackson supposedly offered his enemy a brandy. That fact goes unrecorded. But it is known that Weatherford told him, "I am in your power. Do with me as you please. I am a soldier. I have done the white people all the harm I could; I have fought them bravely: if I had an army, I would yet fight, and contend to the last: but I have none; my people are all gone."

Jackson, who would say that as a young boy he had read the stories of William Wallace, the hero of the Scots, may have seen a touch of Wallace in this Scots-Creek warrior. He admired Weatherford's bravery. And this seemed to be all that was needed to spare Weatherford the hanging Jackson had promised. With a pledge that Red Eagle—William Weatherford—would persuade the rest of his people to surrender, Jackson allowed the defeated Creek leader to leave Fort Jackson. He went to Alabama and took up life as a planter.

Shortly after that extraordinary encounter, Jackson got back to the business at hand: destroying the Creek threat and punishing the Creek without regard for their allegiance. As he wrote to his wife, Rachel, "I will give them, with the permission of heaven, the final stroke."

He summoned the Creek chiefs to a peace council. Under the treaty of Fort Jackson, executed almost one year after the Fort Mims Massacre, the United States confiscated 23 million acres of Creek land, three-fifths of what is now Alabama and one-fifth of Georgia. Jackson also demanded removal of the Indians from lands bordering several states. Even the Indians who had fought beside Jackson had to accept these terms. One of them, a Cherokee chief named Junaluska, reportedly said, "If I had known that Jackson would drive us from our homes, I would have killed him that day at the Horseshoe."[22]

AS THE HERO of Horseshoe Bend, Andrew Jackson was rewarded with a promotion to major general in the U.S. Army. Then he set off for New Orleans with the War of 1812 still raging. The growing legend of Old Hickory was about to reach new heights. On January 8, 1815, Jackson's motley—and multiracial—command included about 4,000 men. Among them were militiamen from neighboring states, pirates, a group of African American freedmen of New Orleans, some Haitians, and Choctaw Indians. Facing an army of British regulars more than twice their number, this American army defeated Major General Sir Edward Michael Pakenham, a brother-in-law of the duke of Wellington, in the greatest land battle of the

War of 1812. The American forces lost thirteen men; British losses were 1,262 wounded, 383 captured, and 291 dead, including Pakenham himself and several other high-ranking officers.

Unfortunately for the men who fought and died that day, the Battle of New Orleans shouldn't have taken place. The war had ended a few days earlier, when the Treaty of Ghent was signed on Christmas Eve, 1814. News traveled slowly in the early nineteenth century. But for Andrew Jackson, the victory burnished his growing personal legend. "When Jackson bragged that he had 'defeated this Boasted army of Lord Wellingtons,' he celebrated the populist triumph of the amateur over the professional, the citizen over the gentleman," the military historians Fred Anderson and Andrew Cayton wrote. "After New Orleans, he seemed as unstoppable as a force of nature."[23]

But even New Orleans was not truly the last engagement of the war. That dubious honor belongs to another incident, which is far more often overlooked. It took place in nearby Florida, which Andrew Jackson continued to eye eagerly. The Spanish were still in control, and Andrew Jackson wanted the territory for America. A few Creek and Seminole Indians who had been allied with Pakenham saw the British depart from American soil in 1815. Their best hope for a victory over the Americans and Andrew Jackson was gone. They retreated to Florida, where Andrew Jackson would continue to press his campaign to clear the territory of Indians and foreign forces. His first opportunity came at a British fort that had been abandoned to a few escaped slaves and their Native American allies.

In 1814, two British officers—Colonel Edward Nicolls of the

Royal Marines and Captain George Woodbine—had established a small stronghold on the Apalachicola River in Florida's western panhandle. For years, escaped slaves had been coming to this territory, which was still in Spain's hands. During the war, more than 300 escaped slaves had taken refuge in the fort, where they were welcomed by the British. By early 1814, Red Stick Creeks escaping from American territory, along with Seminoles and fugitive slaves, had made the fort a headquarters of resistance to Americans. Nicolls reported, "Indians and blacks are very good friends and cooperate bravely together."[24]

When the British evacuated Florida in the spring of 1815, they left this well-constructed fortification in the hands of the freed blacks. As word of the fort spread, it eventually attracted more than 800 fugitive slaves and became known as the "Negro Fort."

In March 1816, Andrew Jackson petitioned the Spanish governor of Florida to destroy the settlement. Jackson was being pressured by slaveholders in Georgia to do something about the Negro Fort, which had become a beacon for runaways. While negotiating with the Spanish authorities, Jackson also instructed Major General Edmund P. Gaines, the man who had arrested Aaron Burr, and was now commander of U.S. military forces in the "Creek nation," to destroy the Negro Fort and "restore the stolen negroes and property to their rightful owners."

On July 27, following a series of skirmishes in which they were routed by the Negro Fort's warriors, the American forces and their 500 Lower Creek allies launched an all-out attack on the fort. The two sides exchanged cannon fire, but the shots of the inexperienced black gunners failed to hit their targets. A shot from the American forces entered the opening to the fort's powder magazine, igniting an explosion that destroyed the fort and its occupants.

The leader of the free blacks—a man named Garson—and a Choctaw chief were among the few who survived the carnage. Both were handed over by the American forces to the Creek allies, who shot Garson and scalped the chief. Other black survivors were returned to slavery.

The destruction of the Negro Fort ended one more attempt at resistance by blacks and Indians in Florida, who increasingly saw their fates as linked. But it did not finish the fighting between white Americans, Florida's Indians, and their new allies: runaway slaves and other free blacks. It would take many years and cost many lives to end that resistance.

Again, Andrew Jackson would be at the heart of the matter. In December 1817, he was directed by President Monroe to lead a campaign against the Creeks and Seminoles in Georgia. He was also charged with preventing Spanish Florida from becoming a refuge for more runaway slaves. Spain, Indians, and escaped slaves: Jackson would attack all three of his most hated enemies at once. With the tacit acknowledgment of the Monroe administration and some ambiguous orders, Andrew Jackson essentially went to war with Spain.

Claiming that there had been an attack on American soil by Seminole warriors, Jackson invaded Florida and captured Pensacola, a Spanish outpost, with barely a shot fired. But he did not stop there. Jackson then arrested two British citizens—Robert Ambrister, a former British Marine; and a seventy-year-old Scottish Indian trader, Alexander Arbuthnot—on the Scotsman's schooner. The two were convicted by a hastily assembled military tribunal of "aiding and abetting the enemy" in what later came to be called the First Seminole War.

In words that presage the controversies over ignoring the Geneva Convention and the legal status of "enemy combatants" in America's "war on terror," Andrew Jackson said, "The laws of war did not apply to conflicts with savages."

The military panel that Jackson had stacked with his loyal officers quickly convicted the men, both British nationals. But when the panel failed to pronounce a death penalty for both men, Jackson stepped in and passed his own sentence. In late April 1818, the two were executed: Ambrister was killed by a firing squad; the elderly Scotsman was hanged from the yardarm of his own schooner.[25]

The case caused a brief but relatively minor diplomatic furor with the British. And Jackson would be reminded of it often in future partisan battles when enemies would taunt him with political ditties like this—

Oh Andy, Oh Andy,
How many men have you killed in your life?
How many weddings to make a wife?

But in the eyes of most Americans, Andrew Jackson was a hero cut from the same cloth as George Washington. And he had accomplished exactly what the Monroe administration tacitly expected. America soon went on to conclude a treaty with Spain in which Florida was turned over to the United States. The long border between Spanish and American possessions, stretching to the Pacific, was also mapped. Andrew Jackson was named the first military governor of the territory. With Florida in hand, Jackson cast an eye toward Cuba, another valuable Spanish possession, but he was reined in by the administration.

Aftermath

The story of the massacre of Indians at Wounded Knee in the late nineteenth century, an episode in which U.S. cavalrymen cut down Sioux Indians, has been made familiar to many Americans. But the deaths at Fort Mims, the killings of the Creeks at Horseshoe Bend, and the many who died at the Negro Fort have mostly disappeared from popular memory and schoolbooks. Andrew Jackson remains, in the view of most Americans, a good if not great president; after all, his grim visage stares out at them from the $20 bill.

That long view overlooks Jackson's record—or at least this early phase of his career. And there are critics who think the view needs to be balanced. "The Creek War of 1813–1814 and Seminole War of 1818 proved to be the most disastrous conflicts in Native American history," concludes Sean Michael O'Brien. "The conflicts shattered the power of the once mighty Muscogee nation forever and paved the way for the removal of all the southeastern tribes from their lands east of the Mississippi."[26]

And it was largely the vision and work of a single man, the man who sat down with Aaron Burr to discuss his invasion plans, but then walked away from his alleged treason, Andrew Jackson had fulfilled a large part of Burr's quest.

Summarizing Jackson's actions in Alabama and Florida, the historian Andrew Burstein writes, "He was much like his predecessors, a man of ambition and enterprise who thought of land as the key to personal and national wealth alike."[27]

But if Andrew Jackson thought that Florida's Indians, free blacks, and "maroons"—free and mixed-race blacks living in established secret communities in the wilds—were finished fighting, both he and the rest of America were seriously mistaken.

III

Madison's Mutiny

1804 Haiti is declared an independent republic after years of bloody slave rebellion.

1807 The British Empire abolishes slave trading.

1808 The importation of slaves and the transatlantic slave trade are prohibited by the United States.

1812 Louisiana is admitted to the Union as a slave state (the eighteenth state).

1816 Indiana is admitted to the Union as a free state (the nineteenth state).

The American Colonization Society is founded in Washington in December, with the goal of resettling emancipated blacks in Africa.

1817 Mississippi is admitted to the Union as a slave state (the twentieth state).

1818 Illinois is admitted to the Union as a free state (the twenty-first state).

1819 Spain cedes East Florida to the United States.

Alabama is admitted to the Union as a slave state (the twenty-second state).

1820 The Missouri Compromise is passed. It allows for admission of Maine as a free state and Missouri as a slave state; under the terms of the compromise, slavery is to be prohibited in any new states that are north of Missouri's southern border. Maine (the twenty-third state) is admitted on March 15; Missouri (the twenty-fourth state) joins the Union on August 10, 1821.

President James Monroe wins reelection in a landslide.

1821 Monroe appoints Andrew Jackson military governor of the Florida Territory.

The American Colonization Society founds Liberia as a haven for freed slaves. Many free African Americans, however, do not wish to emigrate; and by 1870, only 15,000 have immigrated to Africa.

1822 Denmark Vesey's slave rebellion is uncovered and suppressed in Charleston, South Carolina.

1824 In the presidential election, no candidate receives a majority of Electors, although Andrew Jackson clearly wins the popular vote. On February 9, 1825, the House of Representatives decides the election; John Quincy Adams receives enough votes to become the sixth president. When Adams names Henry Clay secretary of state, Jackson angrily denounces them for making a "corrupt bargain." The vote splits the Democratic Republicans into two factions; the Adams-Clay faction will become known as Whigs.

1826 On the fiftieth anniversary of the Declaration of Independence, Thomas Jefferson and John Adams both die. To many, this extraordinary coincidence symbolizes "divine approval" of the United States.

1828 Following a vitriolic campaign, Andrew Jackson is elected the seventh president.

1831 Nat Turner's rebellion terrorizes Virginia.

William Lloyd Garrison publishes the first edition of the abolitionist journal *The Liberator*.

1832 The Anti-Masonic Party holds the first presidential nominating convention in Baltimore, choosing William Wirt, a former Mason.

1833 Slavery is abolished in the British Empire.

1835 Chief Justice John Marshall dies. Andrew Jackson nominates Roger B. Taney to replace him.

1836 The Republic of Texas is founded after breaking away from Mexico.

Arkansas joins the Union as a slave state (the twenty-fifth state).

1837 Michigan joins the Union as a free state (the twenty-sixth state).

On his last day in office, President Jackson recognizes the independence of Texas.

1839 Slaves on the Spanish ship *Amistad* revolt and are brought to America, where they successfully sue for their freedom.

1840 William Henry Harrison, a hero of the Indian wars, defeats incumbent Martin Van Buren to become the ninth president. He dies a month later and is replaced by John Tyler of Virginia, the tenth president and the first to succeed to the office upon the death of a president.

1841 Slaves being transported to New Orleans mutiny aboard the *Creole*.

Led by one Charles Deslondes . . . the insurgents marched on New Orleans. When confronted by United States regulars, they did not break and run but "formed themselves in a line" and returned the fire. . . . Eventually, American soldiers subdued the rebels and hanged and beheaded Deslondes and his confederates. Their mutilated remains hung in public as an object lesson to those who dared to challenge the slave regime.

—IRA BERLIN,
MANY THOUSANDS GONE (1998)[1]

Noble men! Those who have fallen in freedom's conflict, their memories will be cherished by the true-hearted and the God-fearing in all future generations.

—HENRY HIGHLAND GARNET,
BLACK ABOLITIONIST PREACHER, "ADDRESS TO
THE SLAVES OF THE UNITED STATES" (1843)

Insurrection aboard a slave ship did not happen as a spontaneous natural process. It was, rather, the result of calculated human effort—careful communication, detailed planning, precise execution. Every insurrection, regardless of its success, was a remarkable achievement, as the slave ship itself was organized in almost all respects to prevent it.

—MARCUS REDIKER, *THE SLAVE SHIP* (2007)[2]

Atlantic Ocean, near the Bahamas

November 7, 1841

Zephaniah C. Gifford, the first mate of the *Creole*, never saw it coming. How could he? Perhaps he thought that the slaves on board a ship were too ignorant, lazy, or afraid of the lash to fool its officers.

Gifford was standing watch on a balmy autumn night in 1841, surveying the *Creole*'s silent main deck and the soft glow of the moon on the calm Atlantic. Most of the brig's dozen or so crew members and its seven white passengers were asleep as the two-masted ship gently rode the swells about 130 miles northeast of the Bahamas. In one of its cargo holds, the *Creole* carried tobacco destined for the markets in New Orleans. The other holds contained approximately 130 slaves bound for the auction blocks in New Orleans, which had been one of America's most significant ports since it was purchased in 1803.

Suddenly, the stillness of the dark night was broken when Elijah Morris, one of those slaves, bolted from the cargo hold where the male slaves were kept. He shouted to Gifford that one of the other slaves had gone into the women's quarters.

Startled and wary, but not overly alarmed, First Mate Gifford called for help from William Merritt, who was serving as a guard on the *Creole* in exchange for his passage to New Orleans. The male slaves were strictly prohibited from entering the women's quarters, but they were unchained on this voyage. A brig moving slaves from Richmond to New Orleans in the 1840s did not keep its human cargo in the filthy, over-packed, degrading, and deadly conditions that had once been the shame of the transatlantic trade. Wanting healthy, attractive, able-bodied slaves for the New Orleans auction blocks, the *Creole*'s owners and its captain were more interested in the price their cargo would fetch, and seemed unconcerned about security.

Assuming that one of the male slaves was trying to sneak into the female quarters for illicit sex, Gifford and Merritt must have thought they would catch the slaves in the act, give the man a good whipping, and return him to the cargo hold. Gifford sent Merritt into the hold where the women slept.

Lighting his lamp, Merritt was startled to see the face of a slave known as Madison Washington. Washington had been selected as cook for the *Creole*'s cargo of slaves, preparing the twice-a-day ration of hardtack—a brittle, unleavened biscuit—and salted or boiled meat, along with some "coffee" brewed from grain. Addressed as "Doctor" because he could read, Washington was also, according to testimony given later, "a very large and strong slave."

Gifford soon joined Merritt, and the two men attempted to wrestle Washington out of the women's hold. But Washington had other ideas. Jumping through a hatchway, he shouted, "I am going up; I cannot stay here."

The mutiny aboard the *Creole* was under way.

Elijah Morris, who had first raised the alarm as part of a ruse to lure Gifford into the hold and overwhelm him, emerged from the darkness and fired a shot that grazed Gifford's head.

Then Washington shouted to the other slaves in the hold, "We have commenced and must go through; rush, boys, rush. . . . We have got them now." Realizing how frightened many of them were, he called out, "Come up every damned one of you; if you don't lend a hand, I will kill you all and throw you overboard."[3]

Slightly wounded, Gifford raced belowdecks, shouting that a mutiny was under way. Asleep in his stateroom, the *Creole*'s captain, Robert Ensor, roused himself, as seventeen other male slaves rushed out of their hold to join Madison Washington and Elijah Morris. Merritt extinguished the lamp he was holding, but he was still caught by the mutineers, who threatened his life.

John Hewell, a guard and overseer known for his cruelty, was asleep in his berth when he heard the shouts. Grabbing the only musket on board the ship, Hewell went up to the deck. But before he could get off a shot, the musket was wrestled away from him. Because of his reputation, and the misery he was capable of doling out to these slaves as he moved them from slave pen to boat to auction place, Hewell may have already been marked for death by the mutineers.

Captain Ensor emerged from his cabin, carrying a large knife.

He shouted furiously to the other crewmen, trying to rouse them from sleep and rally them to retake the ship. As crewmen joined the fray, a general brawl ensued, and in the close-quarters fighting, Ensor was stabbed several times with his own knife.

"Kill him. Kill the son of a bitch," one of the mutineers cried.

Wounded and bleeding, Ensor clambered into the ship's rigging, where he concealed himself in the darkness, temporarily safe from his pursuers, but unaware of the fate of his wife, four-year-old daughter, and fifteen-year-old niece—all three of them passengers on board the *Creole*.

By this time, the captain's knife had fallen into the hands of two of the mutineers, who chased Hewell, the hated overseer. Hungry for vengeance, the men savagely assaulted Hewell once they caught him. Stabbed more than twenty times, he somehow managed to limp off to his stateroom, where he fell onto a bunk, dying shortly afterward.

In the darkness, with chaos and confusion raging across the decks and holds, the black mutineers had won nearly complete control of the *Creole*. They went from stateroom to stateroom, in search of the other white passengers and crew members. Three men, including a trusted slave who acted as the captain's steward, emerged from one of the staterooms and were spared.

In another stateroom, Theophilus McCargo—the teenage nephew of Thomas McCargo, who owned many of the *Creole*'s slaves—dressed quickly and retrieved two pistols. When the mutineers burst into the room, in search of the wounded Hewell, McCargo fired but missed his mark. When his second pistol misfired, the young man was disarmed and taken captive, but he was spared because of his youth. When the mutineers finally discovered the captain's wife and the two children with her, they too were spared.

HAVING ESCAPED THE onslaught and a near shooting, First Mate Gifford had climbed to the maintop and found Captain Ensor, bleeding and unconscious. Gifford secured the wounded master of the ship to the rigging to prevent him from falling into the sea, thus saving the wounded captain's life.

The violence was nearly over, as the remaining crew and passengers were overwhelmed and taken captive. The mutineers were now well armed with captured weapons. One witness later said that Hewell, the overseer who had been stabbed and died, was decapitated before his body was thrown overboard. A little more than an hour after the mutiny had begun, the *Creole* was in the hands of Madison Washington and his band of followers, fellow captives who were ready to die rather than be forced onto the auction blocks in New Orleans.

Madison Washington—whether that was his actual name or not is uncertain—already had a rather extraordinary tale to tell, having escaped bondage once before and made his way to Canada by the Underground Railroad. Many details of his life story remain sketchy—most accounts of his life and the mutiny were written years after the *Creole* incident, by writers who had never met him.

America's most prominent abolitionist, Frederick Douglass, wrote a highly fictionalized account of Madison Washington and the *Creole* called *The Heroic Slave*, a novella which appeared in 1852, eleven years after the mutiny. Based largely but loosely on the *Cre-*

ole incident, Douglass's tale described how, desperate to rescue his wife from bondage, Madison Washington had left the freedom of Canada and returned to Virginia. Caught while climbing a ladder to her room, Washington was taken to Richmond, held in the slave pens there, and then forced aboard the *Creole* on its fateful voyage to New Orleans.

When the actual fighting aboard the *Creole* was over, Madison Washington told Merritt, the man who had acted as guard in exchange for passage, that he wanted to sail the *Creole* to Liberia, the African state established twenty years earlier by the American Colonization Society as a refuge for emancipated American slaves. At least, that is what the white men who later testified reported. Lacking sufficient water and provisions for such a long voyage, Merritt told Washington that the ship could be taken to the Bahamas instead. Washington certainly would have known that the British had abolished slavery throughout the empire, including the Bahamas, and would free the slaves aboard the *Creole* once the ship had docked.

The *Creole* reached the port of Nassau on November 9, 1841, and the blacks on board were declared free under British law. The *Creole* was released to continue its voyage to New Orleans. Among its passengers were several black women who preferred to proceed back to America, and to slavery in New Orleans.

The seeming ease with which a few dozen slaves could take over a ship is surprising. More surprising still is that the captain and crew of the *Creole* were not more vigilant. In *The Slave Ship*, an account of the barbarity and incredibly ruthless economic efficiency of the slave trade, Marcus Rediker details the threat facing slavers: "Merchants, captains, officers and crew thought about it, worried about it, took practical action against it. Each and all assumed that the enslaved would rise up in a fury and destroy them if given half a chance. For those who ran the slave ship, an insurrection was without a doubt their greatest nightmare. It could extinguish profits and lives in an explosive flash."[4]

But the ease with which the slaves aboard the *Creole* made their insurrection is confirmed in the autobiographical account of Solomon Northup, a free black man from upstate New York who was lured into captivity, drugged, and sold into the slave pens of Washington City (later D.C.) and spent the next twelve years as a slave in Louisiana. Northup's elaborate description of the circumstances aboard the brig *Orleans* probably mirrored the situation on the *Creole* for Washington and his fellow slaves: "After leaving Norfolk [Virginia], the hand-cuffs were taken off, and during the day we were allowed to remain on deck. The Captain selected Robert as his waiter and I was appointed to superintend the cooking department, and the distribution of food and water."

Then Northup describes a conversation with another of the captives:

> *For a long time we talked of our children, our past lives, and of the probabilities of escape. Obtaining possession of the brig was suggested by one of us. We discussed the possibility of our being able, in such an event, to make our way to the harbor of New-York. I knew little of the compass; but the idea of risking the experiment was eagerly*

entertained. The chances, for and against us, in an encounter with the crew, was canvassed. Who could be relied upon, and who could not, the proper time and manner of the attack, were all talked over and over again. From the moment the plot suggested itself I began to hope. I revolved it constantly in my mind. As difficulty after difficulty arose, some ready conceit was at hand, demonstrating how it could be overcome. While others slept, Arthur and I were maturing our plans.[5]

Unfortunately for Northup and Arthur, one of their comrades, named Robert, caught smallpox and died aboard ship. The slaves were quarantined, and so their plan for mutiny died with Robert. Northup was taken to New Orleans, and (as noted above) spent twelve years in slavery before being rescued by friends who sued for his release. His memoir, *Twelve Years a Slave*, was published in 1853.

Two years before the uprising on the *Creole*, Americans had been shocked and divided by a controversy over another slave mutiny, on board the Spanish ship *Amistad*. While sailing from Cuba in June 1839 with a load of illegal slaves, the *Amistad* had been struck by an uprising. The arguments over who owned the ship, the legal status of the Africans aboard, and the very question of slavery's continued existence were going to be brought to the fore as the case wound its way through the American legal system all the way to the Supreme Court. The prominence of the case and the people who would line up on either side of that argument pointed to the sharp, growing divisions that slavery was creating in the United States.

In 1808, under a compromise struck when the Constitution was written twenty years earlier, the foreign slave trade had been outlawed in the United States. It was now a crime to bring slaves into the country, even if slavery itself and the domestic slave trade were still perfectly legal and flourishing. Slavers plying international waters with human cargo from Africa were pirates and criminals, in the eyes of Great Britain, which was aggressively sending the Royal Navy in pursuit of slave ships. But according to American laws, a slaver transporting slaves from one state to another, even in international waters, was a legitimate merchant going about his business.

In the years after the British announced the end of the slave trade in 1807 and then emancipated its slaves in 1833, the Royal Navy had begun an intensive campaign to put an end to the slave trade on the high seas. Illicit slavers sailing under any flag had to keep wary eyes for the ships flying the Union Jack.

By 1839, the work of slavers still bringing slaves from Africa was increasingly dangerous. British antislavery laws were harsh; slavers could be arrested and their ships confiscated. But there were also sizable profits to be made in illegal slaving, as there was a ready market for smuggled slaves in the United States. By reducing the supply of slaves, the end of the slave trade had actually forced up the value of slaves on the auction block, whether they were kidnapped free blacks—as Solomon Northup had been—or illegally captured Africans, like Segbe Pieh.

A native of the British colony of Sierra Leone, Segbe Pieh was kidnapped and taken aboard a Portuguese slaver sometime in late 1838 or early 1839. He had been chained by the neck in a slave coffle, and then marched to the coast, where he was thrown into the hold of the *Tecorah*. Pieh then endured the dreaded Middle Passage from Africa to the Caribbean—weeks chained in a cramped, filthy hold with little to eat and death all around. Aboard the *Tecorah* were some 500 Africans, mostly women and children, chained in pairs, hands to feet, and forced to lie in a deck space less than four feet high. Perhaps one-third of them would not survive the journey; sickness and death rode the slave ships with the human cargo.

Arriving in Cuba, Pieh was spirited—aboard a small boat in the dark of night—into the Cuban jungle, where he would be transformed into a "ladino," a legal, Spanish-speaking slave: that is, a slave born on or brought to the islands before the slave trade was outlawed.

The slaves' first home in the Americas was the barracoon, an oblong hut-like enclosure with no roof. It was both market and prison. David Turnbull, an outspoken British abolitionist, once described the large barracoons built in Havana as a showplace for the slave trade.* One enormous barracoon, located beneath the windows of the Spanish captain general of Cuba, held 1,500 slaves; another held 1,000. A railroad passed by these barracoons, and Spaniards on Cuba took visitors there as a "tourist attraction."[6]

Fed, clothed, their skin rubbed with oil to make it glisten, the slaves were prepared for sale. Pieh and the other contraband slaves were given false identity papers and European names. Although he

*Derived from the same word as "barracks," "barracoon" also entered the American vernacular. Although some people think that the slur "coon" for blacks was derived from the word "raccoon," barracoon is the more likely origin.

and most other Africans with him spoke no Spanish, Segbe Pieh became José Cinque (pronounced "sin-kway"). He could now be sold legally on the slave market.

A Spaniard named José Ruiz and his partner, Pedro Montéz, purchased Cinque for $450 in May 1839. They bought another forty-nine blacks, each slave fetching approximately the same amount, and then four children as well—three girls and a boy, the oldest of whom was nine. From the slave market in Havana, these illegally acquired, but seemingly legal, slaves were taken aboard a small, sleek two-masted Baltimore-built schooner called *Amistad* ("friendship" in Spanish). All of the Spaniards—captain, crew, and Ruiz and Montéz—knew they were breaking the law. But with proper papers for their "legal" slaves, they were unworried.

In June 1839, the *Amistad* set sail from Havana for the Cuban port of Principe. Its fifty-four Africans, all captured by Portuguese in Africa but bearing papers as legal slaves, were bound for another slave market.

But the *Amistad* never reached Principe. On the fourth day out of port, led by Cinque, who used a nail to remove his irons, a group of the captives rose up. With cane-cutting tools stored on board the ship, the mutineers killed the ship's captain and a cook. But Cinque spared the lives of his "owners," Montéz and Ruiz, on their promise to sail the ship back to Africa. For two months, the Spaniards played a game of sailing east toward Africa by day but then sailing north and west under cover of darkness, hoping to eventually reach an American slave state. Wind and currents took them farther north, however, until they reached land off Montauk Point on the eastern coast of Long Island, New York, where Lieutenant Thomas

Gedney of the U.S. Navy brig *Washington* boarded the ship. The two Spaniards immediately informed the American officer that the Africans were murderous pirates.

Gedney had the *Amistad* towed into the port of New London, Connecticut, where the ship and its human cargo were to become part of an elaborate debate and controversy. American abolitionists wanted the men freed; the Spanish government, acting on behalf of Montéz and Ruiz, claimed that the Africans were property and belonged to the Spaniards. The Van Buren administration agreed. But under naval law, Lieutenant Gedney claimed that he had the right of salvage and that the ship was his.

While the African men were held in a prison in New London, Connecticut, abolitionists came to their aid. Having become symbols of the abolitionist cause, the Africans—none of whom could speak English or Spanish—received legal advice from a sympathetic attorney provided and paid for by abolitionist leaders. Suing for their freedom, their attorneys claimed that they were enslaved illegally. To everyone's shock, a local judge agreed. President Van Buren, seeking to dampen southerners' anger and the Spaniards' hostility, pressed his federal attorney to take the matter to the Supreme Court. There, the former president John Quincy Adams, who had returned to the House of Representatives and was an outspoken abolitionist, argued the Africans' case. In an impassioned three-hour speech, Adams insisted that the African captives had never been slaves and that their revolt was similar to that of the American cause in 1776. In a major victory for abolition, the Supreme Court ruled six to one in favor of the Africans. They were freed and eventually returned to Africa, educated in English and converted to Christianity.[7]

LIKE THE MORE famous *Amistad* mutiny, the slave insurrection on board the *Creole* would also end up in an American court. The *Creole* case would ultimately be argued in a New Orleans courtroom and would involve insurance policies. As in the case of the *Amistad*, the *Creole*'s "cargo" was not returned. Both cases created an international diplomatic crisis for the United States. Also, both cases provided ammunition for the increasingly vocal and politically powerful abolitionists of America—as well as their even more powerful and more vocal opponents. Most of all, both cases underscored the great fear that existed in America in the early nineteenth century—the fear of armed blacks fighting to the death for their life, liberty, and pursuit of happiness; murdering whites without mercy; and getting away with it.

THE FEAR OF violent slave insurrections in America was not new. In spite of the ringing words in the Declaration that men had the right to "abolish" the government, those words did not apply to blacks—slave or free. America was deathly afraid of armed black men. This was true in the eighteenth century, when George Washington was reluctant to allow blacks to serve in the Continental Army. It was even truer in the early nineteenth century.

These fears were not imaginary. There had been scattered slave revolts in colonial America, as early as 1711 in New York, which had a large slave population. In 1741, just the fear of a rumored slave revolt had created a panic, and a crushing reign of terror was brought down on New York's slaves. Largely on the basis of unconfirmed rumors of a rebellion, dozens of slaves were tortured and executed. As the historian Jill Lepore of Harvard records in *New York Burning*, "Nearly two hundred slaves were suspected of conspiring to burn every building and murder every white. Tried and convicted before the colony's Supreme Court, thirteen black men were burned at the stake. Seventeen more were hanged, two of their dead bodies chained to posts not far from the Negroes Burial Ground, left to bloat and rot. One jailed man cut his own throat. Another eighty-four men and women were sold into yet more miserable bone-crushing slavery in the Caribbean . . . —'Bonfires of the Negros [*sic*],' one colonist called it."[8]

But localized rumors of slave rebellions and black uprisings were small concerns when set against the reality of what almost every white American—slaveholder or not—feared most. This fear had a name that was well known among many of America's slaves: General Toussaint-Louverture, leader of the slave rebellion on the island of Saint Domingue.

It is quite astonishing that many Americans still grow up never having heard of Toussaint, unaware that Haiti, the nation that became the second independent republic in the western hemisphere, was created by former slaves. As the eighteenth century ended and the nineteenth began, Toussaint's ferocious legacy and the fearsome events on the island that Columbus once named "Hispaniola" had an enormous impact on the future of America.

In December 1791, a wave of rebellion had swept across Saint Domingue, then in French hands.* Inspired by both the American and the French revolutions, this rebellion united Saint Domingue's underclass—the many thousands of black slaves, along with the island's mixed-race mulattoes (from a Spanish word meaning "young mule"). Although technically free, most of Saint Domingue's mulattoes were second-class citizens who joined in the slave revolt against the white rulers. Emerging as their leader was a former coachman named Toussaint who soon adopted the nickname Louverture—"the opening." By 1792, the uprising was so fierce and costly to France that the colony's governor was recalled to Paris and guillotined for his failure to end it.

Saint Domingue's revolution, like France's, was brutal and bloody, fought guerrilla-style, with no mercy shown to civilians. The savagery resulted in the deaths of tens of thousands on both sides, and eventually the exodus of thousands of white refugees fleeing the island for the safety of New Orleans, still in French hands, or elsewhere in the United States. In large part owing to the slave revolt on Saint Domingue, France freed all the slaves in its empire on February 4, 1794, a little-noted outgrowth of the French Revolution and its idealistic call for "Liberté! Egalité! Fraternité!"

The British fretfully watched the carnage on Saint Domingue, from both London and their neighboring West Indies colonies. While happy that their longtime French rivals were suffering a catastrophe, they feared that the virus of slave rebellion might

*The island is today divided between the nations of Haiti and the Dominican Republic.

spread to their other Caribbean possessions, such as Jamaica. When war between Britain and France broke out once more in 1793, the English decided to take advantage of the situation and crush Toussaint's slave rebellion before it could spread. At the same time, they were seizing an opportunity to gain a valuable possession in Saint Domingue, with its rich sugar fields where most of the slaves had been employed in the grueling and deadly work of harvesting cane.

But the British would get far more than they bargained for in the charismatic former coachman. As Adam Hochschild notes in *Bury the Chains*, his history of British emancipation, "Unknown to them, Toussaint L'Ouverture was rapidly turning illiterate rebel slaves into a formidable force. Roughly forty-seven years old when the fighting began, he was described as 'small, frail, very ugly.' Nonetheless, like his similarly short contemporary, Napoleon, he had a powerfully commanding presence. He lived frugally and ate little. Everyone noticed his ever-moving eyes that missed nothing. Perhaps only in Leon Trotsky in the Russian Civil War has history seen another person with no military training or experience so quickly become a leader who could hold great armies at bay."[9]

For the British, the Saint Domingue campaign soon turned into a military disaster of epic proportions. Fighting in the tropical climate in their red wool uniforms, the British soldiers quickly fell prey to the heat and then tropical diseases. Yellow fever swept the ranks. The English suffered greater losses fighting on Saint Domingue than they had earlier in the American Revolution. Finally, they capitulated to the former slave, even establishing trading agreements with Toussaint in exchange for a promise that he would not invade Jamaica.

The proportions of the disaster were truly astonishing. "Of the more than twenty thousand British soldiers sent to St. Domingue during five years of fighting," records Hochschild, "over 60 percent lay buried there. In October 1798, the Union Jack was lowered and Toussaint rode into Port-au-Prince and Cap François—on whose streets he had once driven as liveried coachman."

These horrific casualties were part of the even greater losses the British suffered fighting in the wider West Indies campaign against France. Hochschild writes, "Of the nearly 89,000 white officers and enlisted men who served in the British Army in the West Indies from 1793 to 1801, over 45,000 died in battles or of wounds or disease."[10]

In 1801, following a power struggle within the ranks of the rebel armies on Saint Domingue, Toussaint proclaimed himself a military dictator. He was about to encounter another military dictator. In France, Napoléon Bonaparte had come to power in 1799, made peace with England in 1802, and decided to retake France's former island possession with an eye to making further conquests in North America.

But Napoléon failed to learn from the recent disastrous mistakes of the British. He dispatched his brother-in-law, Charles-Victor-Emanuel Leclerc, with the largest invasion fleet ever to sail out of France. His orders were to "annihilate the government of the blacks of St. Domingue." London quietly acquiesced in Napoléon's invasion plans.

In 1802, the 35,000 French soldiers sent to the island would be greeted by guerrilla warfare and Toussaint's "scorched earth" defense of the island. Across Saint Domingue, drinking wells were polluted with dead animals, and roads were blocked with stones. Some of Napoléon's finest fighting men were astonished at the intense resistance they encountered from the army of former slaves, fighting desperately to retain their hard-won freedom.

Writing from the island, where he would soon succumb to yellow fever, Leclerc warned the emperor, "You will have to exterminate all the blacks in the mountains, women as well as men. Except for the children under twelve. Wipe out half the population of the lowlands, and do not leave the colony a single black who has worn an epaulet. . . . Send 12,000 replacements immediately, and 10 million francs in cash, or St. Domingue is lost forever."

When a long, devastating war of attrition finally reached a standoff, Toussaint agreed to a French offer to negotiate. But it was a ruse. His French opponents simply captured him, threw him into the hold of a ship, and sent him back to France. With Toussaint in chains, hundreds of his officers were similarly swept up and exiled to France. Toussaint died in a French cell on April 7, 1803, ten months after he was taken prisoner.

At the same time, Napoléon reversed the emancipation decree and restored both slavery and the slave trade to the French colonies. The decision reawakened the rebellion on Saint Domingue. The French watched helplessly as appalling numbers of men fell victim to yellow fever; they buried their dead at night in an attempt to disguise their horrific losses. Now led by one of Toussaint's lieutenants, General Jean-Jacques Dessalines, the islanders were fighting to free themselves from enslavement. The French also ratcheted up their cruelty. Rebel prisoners were staked to the ground by the French, and starving dogs were let loose on them. As is often the case, the French atrocities strengthened the resolve of the resistance.

Then, in what was ultimately a catastrophic decision, Napoléon chose to launch a war against England again; the Saint Domingue campaign became a sideshow. By the end of 1803, France had lost more than 50,000 soldiers on the island. These losses were greater than those Napoléon's armies later suffered at Waterloo, his signal defeat in 1815.

On January 1, 1804, Toussaint's successor, Dessalines, proclaimed the republic of Haiti—an Arawak word for "mountain." The French retreated to the other side of the island. Saint Domingue had lost more than half its population, and the country was devastated. But its rebellious former slaves had defeated the two most powerful nations on earth. In one last strike at the hated whites, Dessalines, in an orgy of ruthless recrimination, ordered the massacre of 3,000 of the remaining Frenchmen on Saint Domingue.

Having abandoned hope of retaking the island as a stepping-off point for a campaign in North America, Napoléon decided instead to raise cash for his strapped armies by selling off France's North American territory in Louisiana. Unexpectedly, in 1803, Thomas Jefferson was the recipient of a very valuable gift—made possible by the rebellious slave armies of Saint Domingue.

The events in the Caribbean hardly went unnoticed in America, even before Thomas Jefferson was elected and was able to complete the Louisiana Purchase. The grim spectacle of slaves humiliating and murdering large numbers of their white owners powerfully concentrated the attention of American slave owners. The revolution in Saint Domingue sent shivers through slaveholding America and helped inspire draconian new codes meant to tighten controls over American slaves. Chief among them was a prohibition against teaching slaves to read. American slaveholders thought that keeping slaves illiterate would prevent them from learning about the uprising in the Caribbean. But even illiterate slaves were soon whispering Toussaint's name. And in 1800, even as Jefferson, Burr, and Adams campaigned for the presidency, Richmond, Virginia, was about to get a firsthand taste of what slave insurrection actually meant.

Writing in the *United-States Gazette* in September 1800, a correspondent from Virginia warned:

> *For the week past, we have been under momentary expectation of a rising among the negroes, who have assembled to the number of nine hundred or a thousand. They are armed with desperate weapons and secrete themselves in the woods. God only knows our fate; we have strong guards every night under arms.*[11]

The man responsible for generating that very real fear was a slave who went by the name of Gabriel Prosser. Born in 1776 on a Virginia plantation, Gabriel would be caught up in the feverish air of freedom and democracy then blowing through the country. He and his owner, Thomas Henry Prosser, had practically grown up together. Besides his plantation, Thomas Prosser owned an auction and real estate business, as well as a tavern located on the outskirts of Richmond. Gabriel had become a blacksmith, like his father, and was allowed to hire himself out and keep a portion of his earnings. He was hired out so frequently, in fact, that many believed he was a freedman.

Gabriel possessed another unusual distinction for slaves of his day: he could read. Standing well over six feet tall, and physically imposing, as a man shaped by years of blacksmithing typically was, Gabriel was a proud and impressive figure whose literacy and limited freedom to move about the countryside made him a dangerous black man.

Using his license to move in and about Richmond and its outlying plantations, Gabriel and his two brothers recruited a small army of slave and free blacks willing to revolt against their masters, clearly inspired by the exploits of Toussaint-Louverture. They even had a silk banner inscribed with words that would be familiar to any Virginian of that day: DEATH OR LIBERTY.

Carrying mostly crude weapons, including scythes that had been fashioned into swords, Gabriel's secret army prepared to march on Richmond, home to some 5,700 residents, half of whom were black. Because there were another 4,600 slaves on nearby plantations, whites were actually in the minority in the region. Gabriel planned to take the guns stored in the Richmond armory. The only whites to be spared were Quakers, the French, and others known to oppose slavery.

On August 30, 1800, in the muggy Tidewater heat, Gabriel set out to begin his assault. But before the march on Richmond could get under way, two frightened slaves betrayed him and the "conspiracy" to their owner, who immediately rushed to tell Governor James Monroe of Virginia about the imminent attack. In the midst of campaigning for Thomas Jefferson, Monroe quickly called out the militia and had light artillery dragged into defensive positions around Richmond. A torrential rainstorm that night swamped the rebels' plans. The rebel slave force—estimated variously at between 150 and 1,000 men—never reached the city. With the assistance of the informers, an extensive manhunt was begun the next day. Although Gabriel evaded arrest for a few weeks, he was finally captured on September 25. He was swiftly interrogated, tried, and hanged, along with fifteen others, on October 7. Before being executed, one of the rebels told the court, "I have nothing more to offer than what General Washington would have had to offer had he been taken by the British and put to trial. I have adventured my life in endeavoring to obtain the liberty of my countrymen, and am a willing sacrifice in their cause."[12]

After Gabriel's defeated conspiracy, Virginia and other slave states enacted even harsher new codes that restricted slave meetings on Sundays after work was finished. More important, slave literacy was to come to an end. A little learning for slaves was clearly a dangerous thing. And Virginia issued another rule: that all freed blacks had to leave the state within a year of their emancipation.

Still, the conspiracies and hints of conspiracies continued. As little as two years after the Gabriel rebellion, another slave tried the same thing in Richmond, though again without success.

The next incident that caused tremendous fear and trembling among American slaveholders was also inspired by events in Haiti. It took place in 1811 and became one of the largest slave revolts in

United States history. It started in the area upriver from New Orleans, on the evening of January 8, 1811, with a group of slaves led by a mulatto named Charles—who was from Saint Domingue and whose owner, named Deslondes, had brought him to New Orleans in 1793 following the rebellions on the island. Charles later may have been given his freedom by Deslondes's widow, but he was still working as a paid laborer on the Andry plantation, about thirty-five miles from New Orleans. His emancipation apparently was not enough for Charles. He and his followers overpowered Andry, killed Andry's son, and then began to march toward New Orleans, carrying cane knives, a few guns, and other improvised weapons.

Gathering recruits among slaves on neighboring plantations, and maroons—fugitive slaves who lived in secret hidden communities—the rebellion of Charles Deslondes quickly grew from a small plantation uprising into a major assault on New Orleans. As they marched down the River Road, the insurrectionists began to burn plantations and crops, capturing additional weapons and ammunition as they went. Without any real planning, a rebellion like the one on Haiti seemed to be unfolding in New Orleans. Charles may have hoped to enter New Orleans with an army of freed slaves large enough to capture the arsenal at Fort Saint Charles, expel all whites from the city, and turn New Orleans into a haven for southern slaves.

At the time, New Orleans was one of America's largest cities, with a population of about 25,000, nearly 11,000 of them slaves, and another 6,000 free blacks and other people of color; the city's white population was about 8,000. New Orleans was also the center of trade for American goods coming down the Mississippi and was a crucial port in the valuable trade with the rest of the Caribbean.

Initially met by a small force of local militiamen, Charles's army, now numbering some 200 men, made camp at a farm near New Orleans. By the following day, however, the alarm had been raised. Federal troops and local militia forces confronted the rebel army at the farm and overwhelmed them in a pitched battle. The combined forces, which included members of the New Orleans mulatto militia, killed or captured most of the slaves.

As Ira Berlin recounts in *Many Thousands Gone*, "The insurgents marched on New Orleans. When confronted by United States regulars, they did not break and run but 'formed themselves in a line' and returned the fire. . . . Eventually, American soldiers subdued the rebels and hanged and beheaded Deslondes and his confederates. Their mutilated remains hung in public as an object lesson to those who dared to challenge the slave regime."[13] Most of the 200 slaves who had rebelled were returned to their owners, some in exchange for testimony against Charles and the other leaders.

The rebel slaves of Saint Domingue were also an inspiration to Denmark Vesey, a former slave who had purchased his freedom after winning a lottery. A skilled carpenter, Vesey lived in Charleston, South Carolina, and was a leader in its black church. He had supposedly visited Saint Domingue as a sailor and was also aware of Gabriel's failed conspiracy in Richmond in 1800.

In 1822, Vesey began to plot an insurrection along the same lines as Gabriel's. His forces would strike Charleston's armory, seize its weapons, burn the city, slaughter its white populace, and make their way to Haiti aboard stolen ships. Vesey recruited as many as 9,000 blacks from surrounding plantations. But as the plot built toward its culmination, another slave betrayed Vesey. By June 1822, Vesey was arrested and executed, along with more than thirty of his followers.

Madison Washington and his fellow slaves on the *Creole* were more fortunate. When they sailed into the port of Nassau in the Bahamas, a British colony where slavery had been illegal since 1833, they were welcomed with cheers by Nassau's free black community. The *Creole* "mutineers" were jailed briefly in Nassau, where two of the men died, one from wounds suffered in the fighting aboard the ship, the other from natural causes. The British authorities freed the seventeen remaining men without prosecuting them for either murder or piracy. The chief justice in Nassau told them: "It has pleased God to set you free from the bonds of slavery; may you hereafter lead the lives of good and faithful subjects of Her Majesty's government."[14]

With that, Madison Washington disappeared from history. But his story was told for years to come among the abolitionist forces then beginning to gather strength—and make powerful enemies—in the United States. The most famous escaped slave of the time, the abolitionist leader and internationally renowned lecturer Frederick Douglass, frequently introduced the story of the *Creole* when he spoke. And of course, in 1851, Douglass published *The Heroic Slave*, his fictionalized and highly idealized account of the incident.

But few besides the most committed fringe of abolitionists cheered the British decision or saw these mutineers, or any insurrectionary slaves, as heroic. Slave owners were outraged at the British action. Secretary of State Daniel Webster was forced to pursue the British, even threatening war—though tepidly—over the incident. The case dragged through the courts until 1853, when the British government, following a ruling by an international commission, paid $110,000 to the masters of the *Creole* and its insurers.[15]

In America, the case stirred the forces for and against slavery. On March 21, 1842, while the British still held the men who had mutinied on the *Creole*, and Secretary of State Webster was issuing his threats to the British, an abolitionist, Congressman Joshua Giddings of Ohio, introduced a series of resolutions in the House of Representatives. Asserting that the slaves violated no law in assuming their natural rights, Giddings contended the United States should not try to recover them. Giddings's colleagues attacked him, and the House formally censured him for breaking the "gag rule," which had been used to table any discussion or petitions regarding slavery in the House. Giddings resigned and appealed to his constituents, who immediately reelected him in a landslide.

The small but rising number of abolitionists were still viewed as a "lunatic fringe" in American politics at the beginning of the 1840s. But the situation was changing, as Henry Mayer points out in his biography of the abolitionist firebrand William Lloyd Garrison, publisher of *The Liberator* and perhaps the most uncompromising of American abolitionists: "Giddings' reelection, which Garrison had fully supported in *The Liberator*, suggested that politicians who defied the South might do better at the polls than those who placated it. If such straws in the wind indeed proved indicators of a shift in public opinion, the days of the gag rule were numbered and Congress, instead of avoiding debate on the slavery question, would find itself talking about little else."[16]

Three years later, the "gag rule" was repealed by the House;

this repeal was a powerful portent of the growing strength of the antislavery forces in American politics.

Soon, for America and Americans, the question of slavery would be the only question.

Aftermath

"Slave." "Insurrection." Together, those two words struck fear into the hearts of Americans, before, during, and after independence. Now added to them were the names of Toussaint, Gabriel, Vesey, the *Amistad*, and the *Creole*.

One of the most persistent myths of the slave era is that slaves—both those recently arrived from Africa as well as those born in servitude in America—were docile, somewhat childlike laborers, content with their situation and often treated as "family members" if they were "house slaves." It is a myth that is perpetuated in *Gone with the Wind* and other examples of romanticism about the Deep South, and in the benighted view of such presumably heroic American slaveholders as George Washington, Thomas Jefferson, and Robert E. Lee. Many modern Americans still seem willing to give these American legends a free pass when it comes to the subject of slavery, protesting that they were enlightened owners, who treated their slaves well and sought to emancipate some.

Washington, his admirers love to note, wouldn't sell his slaves because he didn't want to break up families. He treated them well. He emancipated his slaves in his will. But in an earlier time, Washington had offered rewards for the return of runaways. And when he took slaves to New York to serve him as president, they certainly were not free to leave. As for Thomas Jefferson, the author of the Declaration of Independence was totally dependent upon slave labor to operate his plantation, profitably or not. He also contemplated emancipation of his slaves, but was too much in debt to do so at his death. Robert E. Lee was said to be morally opposed to slavery, yet he and the other Lees of Virginia were entrenched members of Virginia's slave aristocracy.

These men *owned* human beings. All the niceties about their feelings and intentions cannot disguise or ameliorate that fact. They had the power of life and death over other human beings, people they could buy or sell at will. And, like many slaveholders, they knew slavery was wrong and an offense to the ideals for which they had fought.

The price Americans were going to pay for slavery was soon to rise in ways that the Founders had never dreamed of. The scattered insurrections and rebellions were about to coalesce as two groups with a common interest—blacks and Native Americans—increasingly joined forces to fight the power of the American government and the slaveholders who continued to press south and west.

The next battlefield would be a familiar one: Florida. When Andrew Jackson first ordered troops into Florida Territory, he couldn't know that this action would be only the first of three linked Seminole Wars, which would last for decades. The Native Americans fighting for survival there had strengthened their ties with their African-Americans allies. United, they began a struggle for survival and freedom that would prove to be one of the longest, most costly wars in American history.

IV

Dade's Promise

TIMELINE

1830 The Indian Removal Act is passed.

1831 Nat Turner leads a slave rebellion in Virginia.

1833 Great Britain outlaws slavery in its colonies. England abolishes slavery in August 1835.

1835 The Second Seminole War begins.

"Dade's Massacre" takes place on December 28.

1836 Congress passes the first "gag rule," automatically tabling any discussion of antislavery measures or petitions. Gag rules remain in effect until 1844.

Martin Van Buren is elected the eighth president.

1837 The Seminole war chief Osceola is captured.

1838 About 14,000 Cherokee Indians from Georgia are removed and herded, on the "Trail of Tears," into Oklahoma. An estimated 4,000 Cherokees die en route.

1842 The Second Seminole War ends.

The Seminole of the present day is a different being from the warlike son of the forest when the tribe was numerous and powerful, and no trouble in the removal of the remnant of the tribe is anticipated.

—*St. Augustine Herald* (May 1835)

Mr. Sheldon and his wife . . . reported the destruction of property between Rosetta and Smyrna, and that Depeysters negroes, with . . . paint on their faces, and Herriot's negroes had gone over to the Indians, and there was reason to believe that a combined operation, and attack would be made upon my small force at Rosetta.

—Major Benjamin Putnam,
Saint Augustine (Florida) Guards, January 4, 1836

This, you may be assured, is a Negro, not an Indian war; and if it be not speedily put down, the South will feel the effects of it on their slave population before the end of next season.[1]

—General Thomas Jesup,
December 1836

King's Road, Florida

December 28, 1835

As a light rain fell at predawn on this December day, the men in Major Francis Dade's company finished their coffee, hardtack, and bacon. Although they were in tropical Florida, the men had felt the change in the air during the night. As they prepared to break camp and resume their march, Major Dade called out to his infantrymen, hoping to keep up their spirits.

"We have now got through all danger; keep up good heart, and when we get to Fort King, I'll give you three days for Christmas!"[2]

It was already three days after Christmas when Major Francis L. Dade made that cheery promise. After four days of travel through what he considered the most dangerous of Florida's Seminole country, where an ambush would be most likely, Dade had let down his guard. A veteran of the sporadic fighting with Seminoles since the United States took control of Florida in 1821, Dade now commanded a relief column, marching from Fort Brooke, near Tampa, to Fort King, near what is now Ocala. His command included seven other officers, a surgeon among them, and 100 infantrymen, many of them recent immigrants to America who had found the army their only real opportunity for employment.

On December 23, Dade and the company left Fort Brooke on a journey of approximately 100 miles. Dade and his men were traveling a rough-hewn "highway" built in 1828. Hacked from the dense wilderness of pinewoods and palmettos, it linked forts Brooke and King. Both had been constructed by the United States to secure Florida after it was "purchased" from Spain in 1821, following the First Seminole War. In spite of the American flag that now flew over Florida, the terrain itself had not changed much since the Spanish conquistadors arrived in the 1500s and met their match in Florida's mosquito-infested swamps and fierce Native resistance.

Juan Ponce de Léon had died in Florida after a deadly encounter with a poisoned Indian arrow in 1521. In 1527, Pánfilo de Narváez lost half of his 600 men in Florida and most of the other half in the Gulf of Mexico when they attempted to sail away in crude boats. And in 1539, the conquistador Hernando de Soto lost his life, along with hundreds of his men, in an attempt to colonize Florida. Its climate, tropical diseases, and hostile Natives had done in these conquistadors and their Spanish armies, with their persistent dreams of cities of gold. Spain's loose colonial control over Florida had come only when they chose instead to establish a "mission system" to convert the Indians rather than trying to conquer them.

Dade's men wore their heavy coats, as the midwinter weather

was growing chilly, even for coastal Florida. Confident that they had passed all danger, Dade had permitted the soldiers to carry their weapons under their greatcoats, to keep the weapons dry. Breaking camp, he divided his command into an advance party, a main body, and a rear guard, marching in two columns. Secure in his belief that danger had passed, he chose not to send out any riders as "flankers" or scouts. In addition to their personal arms, his troops carried with them one cannon, a six-pounder, hauled by a team of oxen.

As morning broke on December 28, the advance guard began the march, a sergeant and five men in single file. After they had moved out 200 yards, the drummer began the beat to "march" and the main column followed in double file. The line of march stretched longer and longer, followed by horses, limber—the small, two-wheeled cart on which the tail of the gun rested—artillery piece, and crew. Then came the lumbering oxen and their wagon with two drivers, then another interval, and at last the rear guard. The woods were silent except for the noise of woodpeckers and squirrels.[3]

Dade was in his early forties, a native Virginian with considerable experience fighting the Seminoles. He had assumed command of this relief force from a fellow officer, Captain George Gardner. First in his class at West Point (1814) and nicknamed "the war god," Gardner had stepped aside to tend his ailing wife, but then rejoined Dade's column.

Slowed by the plodding oxen, the column had made scant progress on the march north; they probably had several more days of travel before them. Since leaving the fort, Dade had seen evidence of Indians along the way, but he did not know that the Seminoles had been following his troopers the entire length of the journey.

Nor could he know that the Seminoles had probably received detailed information about the size of the column and its departure date from Louis Pacheco, a black slave in his forties who traveled with the column as guide and interpreter. Pacheco was a useful man who spoke four languages, including Seminole.

Pacheco knew that the Seminoles and their black allies lay in wait in the tall palmetto grass and amid the forest of pines. As the column approached a small pond along the road, an Indian known to the white men as Jumper gave the war whoop. A storm of fire swept the column, and as John K. Mahon described it, "The sky blue uniforms of the soldiers made easy targets."[4] Dade and the other mounted officers at the front of the column fell instantly. So did nearly half the command. Louis Pacheco dropped to the ground and pretended to be dead.

Jumper's real name was Ote Emathla, and he was brother-in-law of the head Seminole chief, Micanopy. A Red Stick Creek, Jumper had fought against Andrew Jackson in 1818 and had become the chief's "sense bearer," or counselor. Urged on by his warriors, Micanopy had somewhat reluctantly fired the first shot. Halpatter Tustenuggee—whom white men knew as Alligator—would later report that 180 Native Americans and blacks took part in this attack.

Despite the sudden shock of the attack, and the loss of most of their officers, including Major Dade, the soldiers quickly organized a defensive position. The surviving officers, in command of a gun crew, unlimbered the lone field piece and started to fire at the Seminoles. Other soldiers felled large pine trees and started to construct a simple triangular barricade. But their situation was hopeless.

"The soldiers resisted bravely and manned their single artillery piece, firing as rapidly as they could," recorded the historian Joe Knetsch. "As the men attempted to reload the piece, they were shot down by their attackers. In the end, six men manned a makeshift breastwork and fired at the Indians until overwhelmed. . . . The Indian war that all had feared was now on in earnest."[5]

Only three U.S. soldiers, left for dead, escaped the massacre and could make their return to Tampa. All were severely wounded, and only one ultimately survived to provide an account of the attack. In the sudden and sweeping ambush, the Seminoles and their black allies had lost only three men, with another five wounded. When the fighting was over, they moved in quickly and took food, clothing, and ammunition from the fallen soldiers' bodies, before tossing the cannon into the pond and disappearing into the dense Florida woods.

A little later that same day, a second deadly ambush in Florida caught another group of Americans by surprise. Some fifty miles away at Fort King, the intended destination of Dade's column, another group of Indians lay in wait, ready to strike a murderous blow in vengeance. Their target was Wiley Thompson, the federal Indian agent in Florida, who was dining that evening with some of the army officers at Fort King. Thompson was awaiting the arrival of Major Dade and his men. Their task was to complete the "removal" of some Seminoles who had agreed to leave Florida for western territory they had been promised by the federal government.

After finishing his dinner, Thompson took a walk outside the compound with one of the officers. He did not know that he was being watched by a Seminole warrior named Osceola, concealed with other warriors just outside Fort King. The fiercest leader of the Seminole nation, Osceola (or Asi Yolah) was burning for revenge against Thompson, whom he hated for pressing the Seminoles to leave Florida. Catching the two Americans unaware, Osceola's band ambushed, killed, and scalped them and then attacked the fort and killed and scalped four more white soldiers before the alarm was sounded.

When Osceola's war party returned to their camp, some of the warriors were wearing the white men's bloody scalps on their heads; they then hung the scalps from a pole in the middle of their camp to mock the hated Indian agent. Sometime later, the other victorious raiding party returned to the Seminole village, flush with the killing spree that had wiped out nearly the entire relief column in what the rest of America would soon call "Dade's Massacre."

Until this moment, Florida had been as remote to most Americans as the farside of the moon—a distant, exotic land of swamps, snakes, insects, and Indians. Since the Spanish had controlled it for most of America's history, the territory of Florida had mostly been an irritant and source of trouble, especially to southerners in the bordering states. Indians from Florida occasionally raided American settlements and then returned to what was a foreign country. Runaway slaves found a refuge there, either with Jesuits or with Indians. When Andrew Jackson invaded Florida and the Spanish surrendered it, most Americans were content to cheer Jackson and ignore the territory.

All that changed overnight with these Seminole attacks. Up and down the eastern seaboard, American newspapers screamed one word: "Massacre." By the spring of 1836, the Seminoles had also begun to attack plantations, killing white settlers across the state. A widely published etching depicted the killings of Dade's column, Wiley Thompson, and white settlers in Florida. Its headline read: "Massacre of the Whites by the Indians and Blacks of Florida." Beneath the illustration, the caption read, "The above is intended to represent the horrid Massacre of the Whites in Florida in December 1835 and January, February, March and April 1836, when near Four Hundred (including women and children) fell victim to the barbarity of the Negroes and Indians."[6]

Shortly after the treaty that made Florida an American territory in 1821, President James Monroe made the official policy of the United States government very clear: the Seminole Indians "should be removed . . . or concentrated within narrower limits."[7]

Most Americans learn something about the later Monroe Doctrine, a declaration, made in 1823, that Europe's powers should no longer meddle in the affairs of either North or South America. Less familiar is this other "Monroe doctrine," dealing with America's Natives. In modern parlance, Monroe's announced policy was, in essence, "ethnic cleansing"—the deliberate, forced relocation or eradication of a group—and it would soon become an official American policy known as Indian "removal." That brutality, war, plunder, and deception would be used to accomplish the goal was nothing new in the annals of the European and American encounters with Native Americans—from the moment Columbus stepped onto the sands of San Salvador in the fateful fall of 1492, the history of Europeans' relations with the natives they encountered would be written in blood. It was a story of endless betrayals, butchery, and broken promises, from Columbus and the conquistadors through John Smith, the

Bay Colony, the French and Indian War, the War of 1812, and the Creek War. From the outset, "guns, germs, and steel," in Jared M. Diamond's memorable phrase, determined the outcome; superior weapons, the force of numbers, treachery, and disease—the most deadly weapon of all—had been part of a tragic collision between two cultures that surely ranks as one of the saddest and cruelest episodes in history.

That policy perfectly suited Andrew Jackson during his service as Florida's first military governor and throughout the rest of his political career. Elected president for the first of two terms in 1828, the brash, uncompromising hero of New Orleans stood atop a political platform that was fairly simple: suspicion of the upper classes and big business, typified by the Bank of the United States, which Jackson vetoed in 1832; increased voting rights (for white men, at least); a general opening of the political process to the middle and lower classes; freedom of economic opportunity, which included maintaining the role of slavery in America; and eliminating the Indians to open up their lands for white expansion.

Hollywood has left the impression that the great Indian wars took place in the Old West during the late 1800s, a period that many think of as the "cowboy and Indian" days when America's buffalo herds were slaughtered nearly to extinction and Custer made his "last stand." But in fact that was a mopping-up effort. By then, the vast majority of Native nations were nearly wiped out, and the survivors' subjugation was complete. The killing, enslavement, and land theft had begun with the arrival of the first Europeans and was elevated to federal policy with the enforced relocation of Indians from their traditional lands for reservations west of the Mississippi.

Following Monroe's 1821 pronouncement, under subsequent administrations, federal actions toward Indians moved from the broadly popular anti-Indian sentiment and sporadic regional battles to an official federal policy enacted by Congress and Jackson in 1830 and continued by his successor, Martin Van Buren. The tidy word given to this policy was "removal." The Indians called it the Trail of Tears.

In fact, the "Indian question" had been an issue of federal policy from Washington's time. His secretary of war, Henry Knox, signed one of the first formal treaties in 1790, an agreement made with Creek chief Alexander McGillivray, son of a Scottish trader and a French-Indian mother. In 1804, Thomas Jefferson took up the issue in his second inaugural address:

> *The aboriginal inhabitants of these countries I have regarded with the commiseration their history inspires. Endowed with the faculties and occupying a country which left them no desire but to be undisturbed, the stream of overflowing population from other regions directed itself on these shores; without power to divert or habits to contend against it, they have been overwhelmed by the current or driven before it; now reduced within limits too narrow for the hunter's state, humanity enjoins us to teach them agriculture and the domestic arts; to encourage them to that industry which alone can enable them to maintain their place in existence and to prepare them in time for that state of society which to bodily comforts adds the improvement of the mind and morals. We have therefore liberally furnished them with the implements of husbandry and household use; we have placed among them instructors in the arts of first necessity, and they are covered with the aegis of the law against aggressors from among ourselves.*

One of the secrets making up America's hidden history is that many of these "aborigines" had accepted Jefferson's call to "husbandry." In the southeastern United States, in particular, several Native nations had adapted to Anglo ways. Clearly that was not enough for successive administrations, including Andrew Jackson's.

Some historians have attempted to ascribe humane motives to Jackson's call for the wholesale forced migration of Indians from the southeastern states to unsettled lands across the Mississippi. Far better to move them, argued Jackson, than to slaughter them; and slaughter was indeed already happening. In 1831, for instance, Sac tribes under Black Hawk balked at leaving their ancestral lands in Illinois. When a group of some 1,000 Indians actually attempted to surrender to the militia and the regular army, they were cut off by the Mississippi River and then cut down by bayonets and rifle fire. Only about 150 survived.*

In his speech of surrender, recorded on August 27, 1832, at Prairie du Chien, Wisconsin, Black Hawk might have been speaking for millions of Native Americans:

> *Black Hawk has done nothing for which an Indian should be ashamed. He has fought for his countrymen, the squaws and papooses, against white men, who came year after year, to cheat them and take away their lands. You know the cause of our making war. It is known to all white men. They ought to be ashamed of it.*

*Both Abraham Lincoln and Jefferson Davis (later president of the Confederacy) served during the Black Hawk War.

• • •

BUT JUST AS 9/11 and the "war on terror," dominated the American political landscape and decision making in the eight years of the George W. Bush administration, the national security aspect of the "native problem" dictated policy in Jackson's day. Jon Meacham writes in *American Lion*, "To him, the tribes represented the threat of violence, either by their own hands or in alliance with America's foes. When Indians killed white settlers, Jackson tended to see England (or Spain) as the guiding force, believing Indian skirmishes and clashes meant the Indians had been 'excited to war by the secret agents of Great Britain.' "[8]

Jackson wrote to James Monroe, who was then secretary of state, as early as 1817, "The sooner these lands are brought to market, [the sooner] a permanent security will be given to what I deem the most important, as well as the most valuable, part of the nation. This country once settled, our fortification of defense in the lower country, all Europe will cease to look at it with an eye to conquest."[9]

The fact that the territory in question included some of the richest cotton-growing soil in the Southeast added to Jackson's desire to see that "these lands are brought to market."

Passage of the Indian Removal Act in 1830 ensured it. During the Jackson administration, the removals were concentrated on the Five Civilized Tribes of the Southeast. Contrary to popular sentiment of the day and to continuing misrepresentation by historians, the Choctaw, Chickasaw, Creek, Cherokee, and Seminole tribes had by then developed societies that were not only compatible with white culture, but even emulated European styles in some respects. The problem was, to repeat, that their tribal lands happened to be valuable cotton-growing areas. Between 1831 and 1833 the first of the "removals" forced some 4,000 Choctaw from Mississippi into the territory west of Arkansas. During the winter migration, food was scarce and shelter poor. Pneumonia took its toll, and with the summer came cholera, killing the Choctaw by the hundreds.

One of the most eloquent witnesses to the cruelties against the Indians was a young French magistrate studying America's penal system. Observing a Choctaw tribe—which included the old, the sick, the wounded, and newborns—forced to cross the ice-choked Mississippi River during the harsh winter, Alexis de Tocqueville wrote in *Democracy in America*, "In the whole scene, there was an air of destruction, something which betrayed a final and irrevocable adieu; one couldn't watch without feeling one's heart wrung." The Indians, Tocqueville added, "have no longer a country, and soon will not be a people."

The Choctaw were followed by the Chickasaw and then the Creeks, or Muskogee, who did not go as peacefully. The Creek tribe refused to leave, and another Creek War was fought in 1836–1837. Winfield Scott, the American commander of the operation, eventually captured 14,500 Creeks—putting 2,500 of them in chains and marching them all to Oklahoma.

The final series of removals began in 1835, when the Cherokee, who lived in Georgia, became the target. The Cherokee had been among the most "successful" in assimilating European customs. They built roads, schools, and churches; had adopted a system of representational government; and were becoming farmers and cattle ranchers. A warrior named Sequoya had also developed a written

Cherokee language. Establishing a legal, independent Cherokee nation, the Cherokee attempted to resist removal by challenging the policy in the Supreme Court. In one of his final decisions as chief justice, John Marshall ruled in 1831 (*Cherokee Nation v. State of Georgia*) that the Cherokee technically could not bring suit as a "foreign nation" because they were a "Domestic dependent nation": "Their relationship to the United States resembles that of a ward to his guardian."

The Indians had, in other words, tried to fight and switch. But neither strategy worked. They were confronting an irresistible tide of history and an irresistible force in Andrew Jackson. They would disappear, in the Shawnee leader Tecumseh's memorable phrase a generation earlier, as "snow before a summer." In 1838, after Andrew "Sharp Knife" Jackson left office, the United States government finally forced out the 15,000 to 17,000 Cherokees who were still in Georgia. About 4,000 of them died along the Trail of Tears, which took them through Tennessee and Kentucky, across the Ohio and Missouri rivers, and into what would later become (after more broken treaties) the state of Oklahoma.

The strongest resistance to removal came in Florida, where Jackson had already battled the Seminoles. Here, the Indians would carry on a long and costly war, in which thousands of U.S. soldiers and Indians died and millions of dollars were spent, out of a scanty U.S. Treasury. The war began during Jackson's administration but continued well past it. Still, in many ways the Second Seminole War was always Andrew Jackson's war.

An offshoot of the Creek nation, the Muskogee-speaking Seminoles had begun to emigrate into northern Florida in the early

1700s, from the neighboring colonial American territories in Georgia, South Carolina, and other future southern states. Pressed by the expansion of white settlers into their lands, they were also welcomed by the Spanish, who hoped the Natives would act as a buffer against the English (and later American) expansion. The Spanish were also actively encouraging runaway American slaves to escape into Florida.

Settling in Florida, which had been largely emptied of its original Native inhabitants, these transplanted Seminoles established towns like those of the Creeks and maintained Creek societal practices and organization. At first they relied on farming, deer hunting, fishing, and foraging; but they eventually raised livestock and traded with Europeans. Like the Creeks, the Seminoles had essentially done everything that the American administrations from the time of Washington and Jefferson had asked of them.

But during the war of 1812, they aided the British. And after the war, they continued to battle against white settlements until Andrew Jackson used these hostilities as the pretext to invade Florida. When he became president, "Sharp Knife" would be uncompromising in the campaign to rid the Southeast of Indians.

On February 16, 1835, President Andrew Jackson wrote a letter to be read to an assembly of 150 Seminole headsmen and warriors:

> *My Children—I am sorry to have heard that you have been listening to bad counsel. You know me and you know that I would not deceive. The white people are settling around you. The game has disappeared from your country. Your people are poor and hungry. But I have also directed that one third of your people, as provided for*

in the treaty, be removed during the present season. If you listen to the voice of friendship and truth, you will go quietly and voluntarily. But should you listen to the bad birds that are always flying about you, and refuse to move, I have then directed the commanding officer to remove you by force. This will be done, I pray to the Great Spirit therefore to incline you to do what is right."[10]

Aside from the benefits for national defense that Jackson believed removal of the Seminoles would bring to America, and the rich land that would be opened for settlement and for growing cotton, there was another deeply pervasive worry about security, which Jackson did not specifically outline. This was the desperate fear of slave insurrection, especially if runaway slaves and free blacks continued to form alliances with the Indians.

Nothing troubled the sleep of southerners, whether they held slaves or not, more than the idea of a slave revolt. The history of uprisings throughout the colonies and in the early republic was well recorded. The example of Haiti hung over the American South like a storm cloud. "Southern whites were plagued with anxieties," writes the historian Stephen B. Oates. "Can it happen here? What if it happens here?"[11]

In August 1831, three decades after the rebellion in Haiti, it nearly did happen, when Nat Turner led the bloodiest of America's slave insurrections. Nat Turner's rebellion took place in Southampton County, Virginia, about seventy miles south of Richmond, near the North Carolina border. It ultimately failed, as every other major insurrection had, but this uprising transformed the country.

Born on a Virginia plantation in 1800, Nat Turner seemed to

be marked, almost from birth, for an unusual life. Having received a rudimentary education—virtually unheard of for a slave in his day—he had become intensely religious. According to accounts of Turner that later emerged, he had developed what the historian David S. Reynolds describes as a "messianic image of himself as a liberator of his race." Reynolds adds: "He had visions of black and white angels fighting in the heavens, and saw blood on corn—portents, he believed, of his forthcoming victory over his white oppressors."[12]

As a mystic and preacher, Turner used his visions and biblical authority to attract a devoted following among slaves who were permitted to gather for their own worship services on Sunday afternoons after they had been allowed to attend a service at a "white" church. Slaves soon flocked to hear Turner's sermons, dazzled by his visions and his electrifying words.

In his account of Turner's rebellion, Oates describes the centrality of these services for the slave community: "There can be no doubt that the slave church (now a forest clearing, now a tumbledown shack) nourished young Nat's self-esteem and longing for independence. For the slave church was not only a center for underground slave plottings against the master class, but the focal point for an entire subterranean culture the blacks sought to construct beyond the white man's control. The church was both opiate and inspiration, a place where the slaves, through their ring-shout responses and their powerful and unique spirituals, could both escape their lot and protest against it. Here they could find comfort and courage in a black man's God."[13]

Late in August 1831, Turner and about seventy followers started their rampage. Turner believed that a solar eclipse at the time was the portent he had been awaiting—a heavenly sign that the hour of rebellion had come. After killing his own master and the master's family with axes, Turner and his followers set off on a deadly march that spared no one. The killing spree ultimately left fifty-five white men, women, and children dead. And the white folks around Southampton, Virginia, were thrown into utter panic, many of them even fleeing the state.

When Turner's small army was met by a large contingent of white militiamen, Turner attempted a counterattack. But his undisciplined band of freed slaves was vastly outnumbered and outgunned. Following the rout of his forces, Turner escaped and went into hiding. Troops and vigilantes flowed into Virginia, and thousands of soldiers searched the countryside for this lone man who had thrown the region into hysterical terror. Bent on avenging the deaths of the whites, some of the vigilantes simply murdered any blacks they came across; hundreds were cut down in the aftermath of the uprising.

Turner eluded capture for nearly two months, during which he became a frightening bogeyman to the people of the South. On September 17, a reward of $500 was offered for his capture. To whites and blacks alike, Nat Turner had become larger than life, taking on an even greater mystique during his brief disappearance.

In the end, two black men out hunting spotted him and alerted their masters that they had seen the man known as "the Prophet." Following his capture on October 30, 1831, Turner sat in his jail cell for a series of interviews with a sixty-year-old white attorney named Thomas Gray who had ostensibly been assigned to defend the black rebels. Gray asked him, "Do you not find yourself mistaken now?'

Turner replied, "Was not Christ crucified?"*[14]

Even after Turner's execution by hanging, slave owners feared the charismatic Prophet's influence. As the historian Kenneth M. Stampp once wrote, the Turner uprising was an "event which produced mass hysteria from which the whites had not recovered three decades later. The danger that other Nat Turners might emerge, that an even more serious insurrection might some day occur, became an enduring concern as long as the peculiar institution survived."[15]

Stringent new slave laws were passed, and strict censorship laws aimed against abolitionist material were enacted with Andrew Jackson's blessing. Jackson once referred to the antislavery materials mailed by a pioneering abolitionist group, the Anti-Slavery Society, as "a wicked plan of exciting the Negroes to insurrection and massacre." At the other end of the spectrum from Jackson, the abolitionist John Brown—who would later mount his own violent insurrection at Harpers Ferry, Virginia, in October 1859—found inspiration in Nat Turner's plan. "If Nat Turner with fifty men could hold a section of Virginia for several weeks," Brown declared, "an evergrowing band of armed blacks and whites could topple slavery in the state and eventually throughout the South."[16]

After Nat Turner's rebellion, the militant defense of slavery took on a whole new meaning. But in Florida, it would be merged with Jackson's crusade to rid the American Southeast of the Seminole Indians.

---

*Gray's interviews were published as *The Confessions of Nat Turner*, the same title used by the novelist William Styron for his fictionalized version of these events, which won a Pulitzer Prize in 1967.

The Seminoles—or *cimarrón*—had embraced the maroons, runaway and escaped slaves who had established secret communities in many parts of the Americas, from Cuba to Martinique, Brazil, and Jamaica, as well as in New Orleans and Florida, where these slaves had always been welcomed as a part of official Spanish policy. When they first migrated to Florida, the earliest Seminoles had also absorbed the remnants of earlier Florida tribes, such as the Yamasee and Apalachiola, who had been all but wiped out by contact with Europeans, through disease and warfare.

But after the American Revolution, the Seminoles also began to incorporate into their communities the growing number of escaped slaves, coming out of Georgia, South Carolina, and later the Louisiana Territory, who sometimes lived in independent communities. After the Red Stick War and the defeat at Horseshoe Creek in 1814, many Upper Creek refugees had also joined the Seminoles.

The acceptance of runaway slaves and other blacks by the Seminoles would set the Second Seminole War apart from many other Indian wars in American history. It was as much an "African-Indian" war as an "Indian war." For the most part, the so-called "Black Seminoles" were descendants of free Africans and slaves who escaped from coastal South Carolina and Georgia into Florida, beginning in the late 1600s. Joining Native American bands then living in Florida, the Seminole tribe emerged as a multiethnic, biracial alliance. In the nineteenth century, their white enemies called them "Seminole Negroes." Their Indian allies called them "Estelusti," or "Black People." Sometimes the Seminoles held these blacks as slaves, but it was a different sort of slavery from American plantation slavery.

In maroon societies, the historian Richard Price points out, "Seminoles and maroons, during their long history of close collaboration and intermarriage, maintained their separate identities more clearly; they fought side by side but in separate companies against the whites, and maroons (even while being 'domestic slaves') served as trusted advisers and counselors of Seminole chiefs."[17]

Perhaps the most notable and influential of these Black Seminoles was a former slave named Abraham. He had been born into slavery in Pensacola and was about forty at the beginning of the Second Seminole War. Abraham may have won his freedom in the War of 1812, when the British offered freedom to slaves who would enlist to fight against the United States. Abraham later showed up in the town of Micanopy, the principal Seminole chief, and his knowledge of both English and the Muskogee Seminole tongue led to his use as an interpreter. Although he was at first Micanopy's slave, Abraham received his freedom after serving as the chief's interpreter during a trip to Washington in 1826. He later married the widow of another Seminole chief, and by the time of the crisis over removals, Abraham had become chief counsel or "sense bearer" to Micanopy. A portrait of Abraham, one of many water color paintings of significant Seminoles that have been preserved, shows him as a large man wearing a traditional Seminole turban. Abraham was said to be a "sensible and Shrewd Negro" and "the most cunning and intelligent Negro we have seen."[18]

Part of his shrewdness may have been playing both sides in this affair. While seemingly loyal to the Seminole chief. Abraham may have had his own agenda when some of the most important Seminole leaders met in 1832 with Andrew Jackson's envoy, James Gadsden, a veteran of the War of 1812 and the Creek War. In return for $15,400 in cash and some blankets and clothing, many of the Seminole chiefs accepted the terms Gadsden offered under the Treaty of Payne's Landing. It stipulated that every Seminole would be out of Florida within three years. By most accounts, Gadsden "browbeat" the Seminoles into signing the treaty, and many of the tribes that had not been represented in the council later repudiated it.

According to the account of Major Ethan Allen Hitchcock, a Vermonter who served in the Seminole Wars, Gadsden had bribed Abraham, who received $200 when the Treaty of Payne's Landing was signed. A grandson of the Revolutionary War hero and noted iconoclast Ethan Allen, Major Hitchcock was also conspicuously—and unusually, for his day—sympathetic to the Native cause. He would be among the men who discovered the remains of Dade's column in February 1836, and he commented in a journal published after the war: "The government is in the wrong, and this is the chief cause of the persevering opposition of the Indians, who have nobly defended their country against our attempt to enforce a fraudulent treaty. The natives used every means to avoid a war, but were forced into it by the tyranny of the government."

IN OCTOBER 1834, the Seminoles were ordered out of Florida. The major chiefs, including Osceola, met in council. Now about thirty-five, Osceola had come from Creek country in Alabama. After the crushing defeats by Jackson in the Creek War, he and his mother had migrated to Florida with the Red Sticks. Although he was not a chief by inheritance, he was recognized as a natural leader. According to some reports, his wife was a black woman.

"My brothers! The White man says I shall go, and he will send people to make me go," Osceola reportedly told the council. "But I have a rifle, and I have some powder and some lead. I say, we must not leave our homes and lands. If any of our people want to go west we won't let them; and I tell them they are our enemies, and we will treat them so, for the great spirit will protect us."[19]

For the Seminoles and Osceola, the last straw came in 1835, after the government decided to increase the pressure. General Wiley Thompson, a Georgia militia officer whom the government had made an Indian agent, called the Seminole leaders to sign another document, by which they would pledge to leave Florida quietly. In a story that has the ring of legend, one of the chiefs strode to the signing table and plunged his hunting knife into the treaty. That rebellious Indian was known as Osceola.[20]

Wiley Thompson had Osceola clapped in chains. In exchange for his freedom, Osceola signed the document, agreeing to bring in Seminoles who would go west. But once released, the Seminole war chief began to organize the resistance.

By the late fall of 1835, Osceola had the Seminole nation on a full war footing. Burning and plundering white plantations, his bands of Seminole warriors sent Florida into a panic. Then Osceola crafted the plan that would lead to the two attacks of December 28. While one force attacked Dade's column, Osceola took his personal vengeance on Wiley Thompson. Days later, on New Year's Eve, Osceola led a large Indian force against a combined force of army regulars and Florida militiamen and routed them at the Withlacoochee River.

Hearing the news, Jackson appointed General Winfield Scott, another veteran of 1812 who had also fought against the Creeks in Alabama, to take over in Florida. Schooled in European battlefield techniques, Scott was outwitted and overmatched by the guerrilla tactics of the Seminoles. In fairly short order, he was recalled to Alabama.

After an inconclusive year of fighting in which the Seminoles struck freely at white settlements around the state, a new general, Thomas Sidney Jesup, a Virginian by birth, was named to lead the American army's Florida campaign in December 1836. A veteran of the War of 1812, during which he had been taken prisoner by the British, Jesup was a career officer who had served for ten years as quartermaster general of the army and had won praise for reorganizing a department that had traditionally been riddled by corruption and inefficiency. Transferred from the Creek theater to fight the Seminoles, Jesup immediately wrote to his superiors in Washington his assessment of the "real war" he was facing: "This, you may be assured, is a Negro, not an Indian war; and if it be not speedily put down, the South will feel the effects of it on their slave population before the end of next season."[21]

But after a year, Jesup, like Scott, had made no headway against the Seminole resistance. So he switched tactics. Perhaps he had read Napoléon's biography—or Toussaint-Louverture's—because he used the same ploy that had worked against Toussaint. In October 1837, Jesup lured Osceola to negotiations under a flag of truce.

The niceties of war meant nothing to Jesup; disregarding the promise of a truce, he had Osceola put in irons. Chained and imprisoned in a fort in Saint Augustine, the betrayed chief fell victim to malaria three months later. Just before Osceola's death, the famous painter George Catlin visited him and produced a sympathetic portrait showing a noble leader. Even among those who had no love of Indians and supported Indian removal, the deceit that led to Osceola's capture was deemed an act of cowardice. To add to the indignity inflicted on Osceola, one of the doctors who had treated him decapitated his corpse and kept the head, either as a souvenir or as a medical curiosity.

Jesup did not profit from his success in capturing Osceola. The general outpouring of sympathy that followed Osceola's death in captivity left a scar on his record from which Jesup never recovered. Jesup's appointment had come just as a great change swept over Washington, D.C. The man who had dominated American politics almost since his victory over the British in 1815 completed his second term in March 1837. Honoring the tradition established by George Washington, Andrew Jackson did not run again and left the White House. Mounting their first presidential campaign in 1836, the newly formed Whigs failed to coalesce behind a single nominee, sending out three favorite sons instead. The most successful was William Henry Harrison, the former general whose nickname, "Old Tippecanoe," celebrated his reputation as an Indian fighter. Hugh White and Daniel Webster, the other Whigs, finished far behind.

Easily outdistancing them all was Martin Van Buren, Jackson's vice president and handpicked successor, who was to have the distinction of being the first president born an American citizen. An adept tactician tutored by Aaron Burr, Van Buren had begun to master the new politics of group voting, or machine politics, and was responsible for delivering New York's electoral votes to Jackson. But he utterly lacked Jackson's ability to win popular support. Van Buren came to office just as America experienced one of the worst economic downturns in its early history, the Panic of 1837. It would ultimately cost him a second term.

Along with a bad economy, Van Buren had inherited Andrew Jackson's Indian removal policies and continued the federal effort to uproot the remaining Seminoles. But the Seminole War was a quagmire for the Van Buren administration. In yet another change of command, Van Buren dispatched Colonel Zachary Taylor to Florida. Taylor was also a veteran of the War of 1812, though he had seen mostly garrison duty. An Indian fighter who had earned the nickname "Old Rough and Ready," Taylor was a Virginian whose father had served with George Washington during the Revolution. A future American president, he claimed Robert E. Lee as a kinsman and James Madison as a cousin. Madison gave him a commission in the army. But just as Americans were growing increasingly weary of this very expensive and seemingly endless war, Taylor had some success. On Christmas Day 1837, he had tracked down and defeated a Seminole force at Lake Okeechobee.

Around the time that Taylor took command, Robert Reid, the new governor of Florida, suggested bringing in Cuban bloodhounds. They had been used for decades in Jamaica to control rebellious slaves and track runaways. News of the decision to use the dogs, which not only were expensive but required special handlers, brought outraged cries from the abolitionist movement. The northern press and the war's opponents in Congress assailed the bloodhound strategy as contemptible. The "peace hounds," as a skeptical press dubbed them, proved to be of little value in Florida's swamps.

For the next two years, the war dragged on, settling into an endless series of skirmishes until it simply ran out of steam in 1842, having produced less than the government's desired results. In 1840, Taylor requested a transfer to the Indian campaign in the Southwest. Leaderless, their numbers dwindling from attrition, the Seminoles were unable to mount any real resistance. And in Congress, there was no longer any patience with a war that limped along at great expense without—in a modern phrase—any light at the end of the tunnel. Exasperation prompted Senator Daniel Webster to rise and complain, "This Florida war has already cost us over twenty million dollars"—four times the cost of buying Florida from Spain.[22]

And still the Second Seminole War sputtered on, costing the nation a total of more than $30 million. It was the longest U.S. war between the American Revolution and the Vietnam War.

According to the military historian John K. Mahon, the price paid in casualties for the lessons learned and for the ground "liberated" was high. The regular army suffered 1,466 deaths, including 328 men killed in action. Deaths in the navy totaled sixty-nine. Less than one-quarter of the losses were from battle; as it is in most wars, disease took the greatest toll.[23]

Aftermath

Most of the Seminoles had been removed from Florida by the end of 1839. But as the war wound down in 1842, the army was reduced to paying a bounty for the capture or killing of a Seminole warrior. By that point, most of the Seminole elders and war chiefs were gone and one of the few remaining warring leaders, a chief named Halleck, refused to surrender or submit. Using Jesup's trick, the army invited Halleck to visit, promising a meal. While Halleck was at Fort King, the rest of his fighters were captured and Halleck was taken prisoner and transported west. As he boarded the army ship, he said, "I have been hunted like a wolf and now I am about to be sent away like a dog."

On May 10, 1842, President Tyler told Congress that the fighting was near an end, and the war was officially declared over on August 14, 1842. In 1849, a small disturbance flared up again. Known as the Third Seminole War, it led to another sixty Indians' being removed to the west.

By 1850, Indian removal had been accomplished.

"Promises, treaties, and assurances of fatherly solicitude and care were in the end, worth nothing. For public consumption and to assuage private consciences, advocates of removal used the language of religion or of paternalism," Jon Meacham concluded in *American Lion*, his biography of Jackson. "Jackson spoke of himself as the Indians' 'Great Father' all the time—and he almost certainly believed what he was saying. He thought he knew best, and he had convinced himself long before that he was acting in the best interests of both the Indians and the white settlers. But the raw fact remains that the American government—and by extension, the American people of the time—wanted the land. So they took it."[24]

With the Indian "problem" solved east of the Mississippi, the quest for land—and the controversy over whether or not slaves would work that land—did not go away. The problem of slavery in the new territories kept moving farther west as Americans continued to stretch their way across the continent. The next battlegrounds would come as Americans moved into another vestige of Spanish colonial America—Mexico.

The battle over the annexation of Texas by the United States—a breakaway republic recognized by Andrew Jackson on his last day in office in March 1837—and the war with Mexico, begun in 1846, were symptoms of a larger frenzy that was sweeping through America like a virus.

In 1845—the year of Andrew Jackson's death—this fervor was given a name. In a contemporary expansionist periodical, *The United States Magazine and Democratic Review*, the journalist John L. O'Sullivan wrote of "the fulfillment of our manifest destiny to overspread the continent allotted by Providence for the free development of our yearly multiplying millions."

O'Sullivan's phrase, "manifest destiny," was quickly adopted by other publications and politicians. It succinctly expressed a vision, or a quasi-religious mission. Behind this vision was some ideological saber rattling. But the greatest motivator was the insatiable appetite for territory, the force that had driven two American generals turned president, George Washington and Andrew Jackson.

The obsessive desire for Americans to control the entire continent from Atlantic to Pacific had become the Holy Grail in America. As each successive generation pushed the fringes of civilization a little farther, this idea took on the passion of a sacred quest. The rapid westward movement of large groups of settlers was spurred by the development of the famous trails to the West. The Santa Fe Trail linked Independence, Missouri, with the Old Spanish Trail to Los Angeles. The Oregon Trail, mapped by trappers and missionaries, went northwest to the Oregon Territory. The Mormon Trail, first traveled in 1847, initially took the controversial religious group and then other settlers from Illinois to Salt Lake City. And in the Southwest, the Oxbow Route, from Missouri west to California, carried mail under a federal contract.

The fact that California, with its great ports, was still part of Mexico, and that England still laid claim to Oregon, only heightened the aggressiveness of the American desire to control it all.

V

Morse's Code

1834 The Ursuline Convent in Somerville, near Boston, is burned to the ground by a Protestant mob in August.

Under the pseudonym "Brutus," Samuel F. B. Morse publishes a series of anti-Catholic articles collected as *The Foreign Conspiracy Against the Liberties of the United States*.

1835 In Brussels, Alexis de Tocqueville publishes *Democracy in America*.

1836 In Philadelphia in January, James Birney publishes the first issue of *The Philanthropist*, another antislavery newspaper.

Elijah Lovejoy, publisher of an abolitionist newspaper, is killed by a mob in Alton, Illinois, in November.

1838 Philadelphia's Pennsylvania Hall, the site of antislavery meetings, is burned to the ground by a pro-slavery mob in May.

1844 The first telegraphic message—"What hath God wrought!"—is sent from Washington, D.C., to Baltimore by Samuel F. B. Morse on May 24.

The deadly anti-Catholic "Bible Riots" sweep Philadelphia during May and July.

The Baptist Church, divided over the question of slavery, splits into northern and southern conventions. The Methodist Church, South, also breaks away over slavery.

James K. Polk is elected the eleventh president, defeating the Whig candidate Henry Clay and the Liberty Party candidate James K. Birney. An abolitionist, Birney may have drawn support away from Clay, especially in New York, possibly costing Clay the election.

1845 Florida is admitted to the Union as a slave state (the twenty-seventh state).

Andrew Jackson dies at the Hermitage, his home near Nashville, Tennessee, on June 8.

Texas is admitted to the Union as a slave state (the twenty-eighth state).

The former slave Frederick Douglass, who escaped to become an abolitionist leader, publishes *The Narrative of the Life of Frederick Douglass*.

The potato blight strikes Ireland, adding 1.5 million more Irish to the waves of immigrants coming to America over the next few years.

1846 War is declared on Mexico on May 13.

The "Bear Flag Revolt." Americans declare a Republic of California in June, breaking free from Mexican rule.

Iowa is admitted to the Union as a free state (the twenty-ninth state).

Surely American Protestants, freemen, have discernment enough to discover beneath them the cloven foot of this subtle foreign heresy. They will see that Popery is now, what it has ever been, a system of the darkest political intrigue and despotism, cloaking itself to avoid attack under the sacred name of religion.

—"BRUTUS" (SAMUEL F. B. MORSE),
FOREIGN CONSPIRACY AGAINST THE LIBERTIES OF THE UNITED STATES (1834)

There was a girl thirteen years old whom I knew in the School, who resided in the neighborhood of my mother, and with whom I had been familiar. She told me one day at school of the conduct of a priest with her at confession, at which I was astonished. It was of so criminal and shameful a nature, I could hardly believe it, and yet I had so much confidence that she spoke the truth, that I could not discredit it.

—MARIA MONK,
AWFUL DISCLOSURES OF MARIA MONK (1836)[1]

As a nation, we began by declaring that "all men are created equal." We now practically read it "all men are created equal, except negroes." When the Know-Nothings get control, it will read "all men are created equal, except negroes, and foreigners and catholics." When it comes to this I should prefer emigrating to some country where they make no pretence of loving liberty—to Russia, for instance, where despotism can be taken pure, and without the base alloy of hypocrisy.

—ABRAHAM LINCOLN (1855)

Philadelphia

May–July 1844

THE BLOODSHED BEGAN over the Bible.

In 1844 Philadelphia, the City of Brotherly Love, was not very brotherly. And there wasn't much love.

On Friday, May 3, 1844, the American Republican Party, an anti-immigrant, anti-Catholic Protestant group also known as the Nativist Party, set up a platform to hold a meeting—a "Save the Bible" rally—in the Third Ward of Kensington, then a predominantly Irish suburb of Philadelphia. The speakers delivered a series of invective-filled tirades directed against the Irish, the pope, the Catholic Church, and immigrants. All of them, including some of Philadelphia's Protestant clergymen, railed that the Germans and Irish "wanted to get the Constitution of the U.S. into their own hands and sell it to a foreign power."

At the heart of their anger was a belief, widely held among

Philadelphia's Protestants, that the city's Catholic bishop, Dublin-born Francis Patrick Kenrick, was trying to remove the Bible from Philadelphia's public schools. These rumors were part of a much more widespread and virulent conspiracy theory—widely held by nineteenth-century Americans—that the pope was planning to take over America.

During the deadly cholera epidemic that swept Philadelphia in 1832, killing more than 1,000 people, Bishop Kenrick had won praise for his unstinting efforts in battling the contagion and tending its victims. Part of a worldwide cholera pandemic, which also struck New York, New Haven, Boston, and other New England cities that year, Philadelphia's outbreak was blamed on immigrants, and the Irish in particular. Since the cause of cholera—water contaminated by feces—was not yet understood and recent Irish immigrants were most likely to live in the crowded, squalid conditions that fostered cholera, it was a short step to blame them for the deadly outbreak.

As was typical of immigrant populations in urban areas, the Irish actually suffered disproportionately during the crisis. A large number of Irish laborers had arrived in Philadelphia to work on construction of the Philadelphia and Columbia Railroad, a pioneering railroad and canal project begun in 1829. In 2009, a mass grave with the remains of fifty-seven Irish railroad workers, probably victims of the epidemic but whose fate was unknown, was discovered near Philadelphia. The section of the old railroad running from the city to western suburbs was called the "Main Line of Public Works," and is still known as the Main Line.

The city's Irish enclave was mushrooming when the cholera

epidemic of 1832 struck and, as noted above, many Philadelphians blamed the outbreak on these Irish immigrants. That idea was reinforced in 1837 by an epidemic of typhus, which was widely called the "Irish disease" and is spread by lice. In his history of the Irish in Philadelphia, Dennis Clark wrote, "The republic was building, and the work to be done strained at the backs of immigrant and native alike. [The Irish were] clannish, politically sensitive and intolerant of interference. In the alleys where they lived, conditions did not breed tolerance. In 1832, a citizens' committee found in a workers' area near the Delaware fifty-five families without a single privy for their use."[2]

Among many Americans of the era, fears about religion, sanitation, politics, and economics were fueled by vitriolic propaganda. Over the following years, anti-Catholic and anti-Irish sentiments in mainstream Protestant America deepened, as the flow of immigrants increased. As the essayist and novelist Peter Quinn wrote, "The Irish were swiftly identified in the popular mind with poverty, disease, alcohol abuse, crime and violence—all the enduring pathologies of the urban poor. Indeed the level of social turmoil that followed the Irish into America's cities would not be seen again for another century, until the massive exodus of African Americans from the rural South to urban North."[3]

On this warm May afternoon in 1844, the crowd of Protestant Nativists was fairly small, possibly numbering 100. When a group of Irish residents—most of them unemployed young locals, just hanging about—jeered and then started to tear down the platform, the Nativist speakers and the crowd retreated.

But three days later, on Monday, May 6, the Nativists returned in force, now numbering some 3,000. When it began to rain, they moved their rally to the nearby "Nanny Goat Market," an Irish marketplace, where the inflammatory remarks continued to flow from the speakers, some of them evangelical Protestant clergymen. The Irish were derided as "scum unloaded on American wharves."

Again, the Irish locals responded to the Protestant speakers' invective with heckling and jeers. According to many accounts of these events, alcohol seemed to flow as freely as the insults. Nineteenth-century Americans were often hard drinkers, and cheap whiskey was more available than fresh water. Words soon became shoves, and fistfights broke out. As crowds of Irish and Nativists pressed toward each other, one man pulled out a pistol. Shots were fired, and soon more weapons were produced. Rocks and clubs were rapidly added to the mix. When one Irishman, Patrick Fisher, attempted to stop a fight, he was shot in the face.

The violence quickly escalated as Irish residents began sniping at the Protestant Nativists from their homes and rooftops. Exposed in the open market, the crowd of Protestants made easy targets. An eighteen-year-old Protestant leather tanner, George Schiffler, was hit and died instantly, the first casualty of the Philadelphia "Bible Riots."

As the fighting spread through the Irish neighborhood, some of the Nativists went looking for reinforcements, and groups of armed Protestants soon began to arrive. Fueled by alcohol, they set about stoning houses and breaking windows and doors. When the sheriff and his deputies arrived to bring the melee under control, they were armed only with clubs and were seriously outnumbered. The sheriff requested help from the state militia but was turned down by its commander, General Cadwalader. By nightfall, Nativist gangs were wreaking havoc all over Kensington, breaking into houses, destroying buildings, and driving Irish families from their homes. Bonfires flickered around the smoldering neighborhood.

The rioting subsided as the evening wore on, slowly dissipating after the long hours of street-by-street fighting. But sporadic gunfire could be heard throughout the night.

Overnight, the Nativist Party produced flyers offering a $1,000 reward for the killers of George Schiffler, the young man who had been the first to fall. Schiffler was going to be turned into a Nativist martyr. The flyers called for an outpouring of Protestant strength, and advised, "Let Every Man Come Prepared to Defend Himself." On the following morning, an extremist Nativist newspaper compared the previous night's action to the St. Bartholomew's Day massacre in France, where thousands of Protestants had been murdered by organized Catholic mobs in 1572: "The bloody hand of the Pope has stretched forth to our destruction. Now we call on our fellow citizens, who regard free institutions, whether they be native or adopted, to arm. Our liberties are now to be fought for—let us not be slack in our preparations."

Spurred on by the propaganda and the deaths of several Protestants, the Nativists returned to Kensington, well armed this time. At least 3,000 strong, they spent the next day and night burning down houses. The roving bands of Nativists carried a tattered American flag with a banner attached that read, "This is the flag that was trampled underfoot by the *Irish Papists*."

Inside the Hibernia Hose House, a volunteer fire brigade and Irish meetinghouse, Catholic defenders awaited the onslaught of the Protestant mob. When the Catholics opened fire, the Protestants returned it, and a full-scale pitched gun battle for Kensington was again under way. Four Protestants were killed in the shooting. Retaliating, the Nativists began to torch the neighborhood, and the old wooden structures were soon blazing.

By the day's end, the fires had destroyed the Hibernia fire station, thirty Irish homes, and a market. The crowd dispersed only after the arrival of a militia force, led by General John Cadwalader, scion of one of Philadelphia's most prominent families and a descendant of a Revolutionary War hero of the same name.* Cadwalader's troops were able to restore order, and Bishop Kenrick issued a statement imploring his Catholic flock not to resort to violence. When the rioting continued on Wednesday, May 8, Protestant Nativists dominated it. After setting fire to more houses, they headed for St. Michael's Catholic Church and its rectory, which were guarded by a small militia detachment. Father Michael Donohoe, the rector, was an outspoken critic of the Protestant religious practices forced on Catholic children in the city's public schools.

Targeted by the mobs, Donohoe had left town earlier in the week. Unappeased, the Nativist mobs set still more fires, broke into

---

*The first John Cadwalader arrived in the Pennsylvania colony with William Penn. His grandson, General John Cadwalader, was a hero of the American Revolution whose epitaph was composed by Thomas Paine: "His early and inflexible patriotism will endear his memory to all true friends of the American Revolution. It may with strictest justice be said of him, that he possessed a heart incapable of deceiving. His manners were formed on the nicest sense of honor and the whole tenor of his life was governed by this principle." That general's grandson, John Cadwalader, who helped suppress the Bible Riots, was also a general, and a prominent attorney.

St. Michael's, destroyed the rectory, and threw Donohoe's entire library into the street. The church building was then set afire, and it burned until its steeple came crashing down, to Nativist cheers. Many of the Protestants were drunk—or at least drinking.

Seeing one Catholic church in flames, the Nativist crowd next marched on the seminary run by the Sisters of Charity, who had won praise during the cholera epidemic for their fearless devotion to tending the sick. Although the nuns had left the building, a housekeeper was hit in the face with a stone when she opened the door. Despite being cautioned to restrain themselves, Irish Catholics guarding a nearby church could hold back no longer. They opened fire on the Protestants, killing one man instantly and fatally wounding another. The seminary and several homes were then set ablaze before soldiers arrived and the fires were finally contained.

Kensington's mayor, John Morin Scott, had stationed troops near St. Augustine's Church, located on Fourth Street between Vine and New Streets, and was trying to restore order. When Scott was hit in the chest with a stone, the rioters went ahead and burned down the building, cheering as another Catholic steeple collapsed. That marked the end of the Kensington riots, with a death toll of at least seven, and more than twenty wounded, two of them fatally. Some 300 Irish Catholics had been forced to flee their homes.

In the following days, Mayor Scott established a force to protect Catholic churches, and Bishop Kenrick ordered all churches to be closed on the approaching Sunday, May 12, to avoid any further provocations. But the violence seemed to have spent itself, and an uneasy calm settled on a smoldering and badly damaged Philadelphia.

With the approach of Independence Day, however, threats of violence began to be whispered once more. The riots in May had strengthened the number and resolve of the Protestant forces. On July 4, thousands of people marched through Philadelphia in a stark show of strength by the growing Native American Party. The parade was pro-American, anti-Irish, and anti-Catholic. What began as a celebration of "Life, Liberty, and the Pursuit of Happiness" in the birthplace of the Declaration descended quickly into a bacchanal of bigotry.

Responding to the rumors of another Nativist rally around the birthday of American independence, the parishioners at St. Philip Neri Church in Southwark, another Catholic neighborhood in Philadelphia, had requested permission from the governor to form a militia and draw twenty muskets from the town arsenal.

Before the parade had stepped off, word got out that the governor had agreed to allow the Irish in Southwark to keep arms on hand for their defense. This news seemed to provoke even more Protestant anger. In an attempt to placate the mob, the sheriff removed the guns from St. Philip Neri on the morning of July 4. But his gesture did nothing to placate the Protestant crowds, and the mood turned even more ugly and threatening. The parade became a Protestant rally, but even so, there was no confrontation.

That would change two days later. On July 6, a mob gathered outside the church, which was now guarded by local militiamen. By the early morning of July 7, the Nativist crowds had intimidated the militiamen and the contingent withdrew, leaving the church unguarded. Disarmed and unprotected, the Irish who had stayed in the building were badly beaten as they evacuated it.

The Nativist crowd then descended on St. Philip Neri. Fires were set, holy pictures were slashed, and other holy objects were desecrated. After some of the rioters had moved into the church, the city militia began to try and clear them from nearby Queen Street. But the passions of the crowd had been pushed to uncontrollable heights. The mob of rioters fought back, shooting at the militiamen with three cannon taken from a ship at the nearby docks. The fighting ended around midnight after General Cadwalader's militia again moved in to quell the violence.

In two separate bouts of rioting in May and July, nearly two dozen Philadelphians, both Protestant and Catholic, were killed, and many more were injured.

Bishop Kenrick hoped that the civil authorities would restore calm and order and that the rioters would be punished, but within days, a grand jury had blamed the rioting on the Irish. Adding insult to injury, the grand jury contended that the outbreak of violence was due to "the efforts of a portion of the community to exclude the Bible from public schools."

The non-Irish, non-Catholic juries acquitted Nativists and convicted Catholics.

Anti-Catholicism had a long and painful history in America, but the seeds of this deadly local conflict had been planted as recently as 1842, when Bishop Kenrick wrote a letter to the board of controllers of public schools, asking that Catholic children be allowed to read from the Douay version of the Bible, a translation widely used by Catholic churches, and also that they be excused from other religious teachings, many of which had a distinct anti-Catholic tone, while at school. In most American schools of this time, the day began with Bible readings, and the only Bible used was the King James Version, the staple of English Protestantism since 1611.

In the months following the first letter to the board of controllers, anti-Catholic groups twisted Kenrick's request into an attack against the Bible and Protestantism. Combined with the growing anti-Catholic press—picture a nineteenth-century version of hateful talk radio and Web sites—that intensified the intolerant mood, the rumors that the Catholic bishop was trying to "take the Bible out of schools" grew into talk of a broader plot against Protestantism. With a struggling economy, which had not fully recovered from the severe economic downturn known as the Panic of 1837, the mood in Philadelphia was dark. Added to the volatile mix of sectarian hatred and financial desperation was the growing division over abolition, a movement in which Philadelphia was emerging as a center. Class, race, and religious differences were fueling anger that completely betrayed the notion of America as a "benevolent empire" of Protestant Christianity.

"NEVER WAS A city more disgraced; never a city more justly punished," wrote a correspondent in Philadelphia's abolitionist newspaper, the *Pennsylvania Freeman*, on July 18, 1844. "Our sins have been their own punishment, and we have been made to eat most bitterly the fruits of our own doings. It is to be hoped that our city will now learn wisdom and put away the evil of her doings, and that she will at last be persuaded to respect the rights of the poorest and most unpopular of her citizens."[4]

Benjamin Lundy, a Quaker, had begun publishing the *Pennsylvania Freeman* as the *National Enquirer* in 1836. In 1838, after Lundy's business and possessions went up in flames at Pennsylvania Hall, an abolitionist meeting place that was burned to the ground by a mob, he left the city. The journal was taken over by another Quaker, the poet and ardent abolitionist John Greenleaf Whittier, who renamed it. When the *Pennsylvania Freeman* wrote of Philadelphia's "sins," it was not referring only to the Bible Riots.

The city had been ripped apart by violence aimed at stilling its increasingly vocal abolitionist faction. A few years before Pennsylvania Hall was burned out, mobs had torched the homes of forty black Philadelphia families. The rising tide of abolitionism and the destructive counterattacks that the movement was beginning to provoke were warning signs that America was fracturing along some very deep fault lines: race, region, party, and politics. Although slavery had lost its hold in most northern states by the beginning of the nineteenth century, there was no love for blacks, and no integrated society, anywhere in America, save among a few radical abolitionists. In New York and New England, anti-abolitionist mobs had begun to attack abolitionist leaders—including William Lloyd Garrison, who was paraded through Boston with a noose around his neck in the same year that Pennsylvania Hall was burned.

An uncompromising abolitionist who preferred disunion to an America with slavery, Garrison had given John Greenleaf Whittier his first jobs as an editor and helped shape the young poet and writer's ardent devotion to the cause. In 1833, Garrison and Whittier were among the founders of the American Anti-Slavery Society, originally formed at the Adelphi Building in Philadelphia but later based in New York. With a surge in evangelical Protestantism, following a second Great Awakening, abolitionism—along with temperance, prison reform, and a budding suffrage movement—spoke to the country's changing mood and ethics. The American Anti-Slavery Society quickly grew to more than 1,300 local chapters around the country and had 250,000 members.

But the public at large did not share the sentiments of these nineteenth-century reformers, whose ideas about the races, about temperance, and about votes for women were far from the mainstream of American public opinion. Mobs frequently struck at abolitionist meetings, attacking the speakers and destroying their printing presses. The antipathy toward "nigger lovers" and the whole gamut of "papists, micks, wops, and dagos" that made up the bulk of arriving immigrants was evidence of a deep, dark mean streak in America about these issues. And it was creating a crisis in the mid-1830s, just as America was beginning to suffer one of the worst economic downturns of the century.

The Panic of 1837 was one of the first broad-scale, far-reaching financial dislocations in American history. Before that time, most Americans were still subsistence farmers, relatively insulated from the worst impact of boom-and-bust business cycles. But the problems facing the nation in 1837 sound all too familiar to twenty-first-century Americans: speculative real estate bubbles, bank failures, unemployment in a rapidly changing economic landscape, and a crash in prices due to oversupply (back then it was cotton prices). Ultimately, a full-blown depression hit America and lasted five years. "The Panic of 1837," writes the historian Daniel Walker Howe, "merged with that of 1839 into a prolonged period of hard times that, in severity, and duration, was exceeded only by the great depression that began ninety years later, in 1929."[5]

This panic began almost as soon as Martin Van Buren took the oath of office, partly because of a series of events that had been happening under his predecessor, Andrew Jackson. But the opposition Whigs successfully blamed "Martin Van Ruin" for the crippling economic disaster and doomed his chances for reelection. Immigrants and free blacks—the least powerful groups in America at the time—also bore the brunt of anger over the loss of jobs and income, feeding the powerful anti-Catholic sentiment that already coursed openly through American society.

The anti-immigrant, anti-Catholic sentiment that sparked the Bible Riots went well beyond the streets of Philadelphia. It ran deep in American waters. The sectarian hatred between Protestants and Catholics had arrived on America's shores with the first Europeans to land. After the European religious wars of the Reformation era; the persecution of Protestants under England's queen, Bloody Mary; and the long history of Spain's attempts to invade England during the reign of Elizabeth I, there was deep mistrust and hatred of Catholicism among many Anglo-Americans. The 1565 massacre of French Huguenots in Florida by the Spanish had been the first bloodletting on America's shores. Centuries of anti-Catholic propaganda had deepened the antagonism. The hatred went beyond intolerance; it was a sectarian blood feud (which, as the sad history of Northern Ireland proves, continued well into the twentieth century).

In some of the American colonies, this deep division was apparent in laws limiting or prohibiting Catholics. In colonial Boston,

Catholic priests were banned; and each November 5—which the English celebrated as Guy Fawkes Day to commemorate a failed Catholic plot against Parliament—was celebrated as Pope's Day, and effigies of the pope were paraded around Boston.

Pennsylvania had seen a wave of immigrant bashing as early as the 1750s, when even the greatest figure of the American Enlightenment, Benjamin Franklin, had voiced anti-immigrant sentiments. "Few of their children in the country learn English," Franklin once complained. "The signs in our streets have inscriptions in both languages. . . . Unless the stream of their importation could be turned . . . they will soon so outnumber us that all the advantages we have will not be able to preserve our language, and even our government will become precarious."

The language so vexing to him was the German spoken by new arrivals to Pennsylvania in the 1750s, a wave Franklin viewed as the "most stupid of their nation." At about the same time, the Lutheran minister Henry Mühlenberg, himself a recent arrival, worried, "The whole country is being flooded with ordinary, extraordinary and unprecedented wickedness and crimes. . . . Oh what a fearful thing it is to have so many thousands of unruly and brazen sinners come into this free air and unfenced country."[6]

Often, disdain for foreigners was inflamed by religion. Boston's Puritans, for example, banned Catholic priests and Quakers—hanging several "Friends" for good measure. But the greatest scorn was generally reserved for Catholics—usually the Irish, French, Spanish, and Italians. Generations of white American Protestants resented newly arriving "papists."

Anti-Catholic sentiment even fueled some of America's revolutionary ardor. After fighting in the French and Indian War, many Americans were incensed when they learned that King George III had accepted the Peace of Paris in 1763 and the Quebec Act of 1774, both allowing Catholic French-Canadians greater freedom of religion. The law was considered not only a betrayal of Protestantism in pre-revolutionary America but also a direct assault on land claims made by American speculators who had fought against the French and now felt deceived and abandoned by England's king and Parliament.

After the Quebec Act was announced, the Congregational minister Ezra Stiles (who would later be president of Yale) complained loudly and bitterly that the enactment had established the "Roman Church and *Idolatry*." New York's John Jay, a delegate to the Continental Congress and later the first chief justice of the United States, spoke for many Americans when he expressed his fear that a wave of Catholic immigration would "reduce the ancient free Protestant colonies to [a] state of slavery." In the Continental Congress, Jay railed "in astonishment" that Parliament should ever consent to establish "a religion that had deluged your island in blood and spread impiety, bigotry, persecution, murder and rebellion throughout every part of the world."

Once independent, the new nation began solidifying these deep-seeded prejudices against immigrants into law. In considering New York's state constitution, for instance, Jay suggested erecting "a wall of brass around the country for the exclusion of Catholics."[7] In Maryland, a supposed haven for "papists," Roman Catholics were forbidden to vote and hold public office.

In 1790, the first federal citizenship law restricted naturalization to "free white persons" who had been in the country for two years. That waiting period was later lengthened to five years and, in 1798, to fourteen years. Then, as now, politics was a factor. Federalists feared that too many immigrants were joining the opposition. Under the 1798 Alien Act—with war in the air, coincidentally—President Adams had license to deport anyone he considered "dangerous." Although his secretary of state favored deportations of numerous foreign-born Americans, Adams never put anybody on a boat.[8] Back then, the French aroused the most suspicion. But a wave of "wild Irish" refugees was thought to include dangerous radicals. The anti-Catholic mood was also on display in the hatred voiced toward Spain and Mexico.

But the great influx of predominantly Irish and German immigrants, most of them Roman Catholics, was responsible for heightening the tension. Although the mass immigration of the Irish began in the 1840s, during the Potato Blight, there had already been a huge influx of Irish and German immigrants before that, beginning after 1815. During the 1830s, 500,000 immigrants settled in New York City alone. Surpassing in numbers the Episcopalians, Methodists, and Congregationalists, Roman Catholics became America's largest religious group. "By 1850 the Roman Catholic Church had become the largest denomination in the country, a status never thereafter surrendered to any other church," write Edwin Gaustad and Leigh Schmidt. "'Anti-popery,' the bread and butter of Protestantism, had reared its head regularly in America, but now it raised itself to new heights."[9]

By the early nineteenth century, dozens of anti-Catholic periodicals were being published in the United States. Many of these argued that Catholics could not be trusted, because they owed their allegiance to the pope instead of to their new country. The famous political cartoonist Thomas Nast repeatedly depicted popish conspiracies to take over the United States. The most outlandish claims—widely accepted—included a prediction that the pope would land with a papal army and set up a new Vatican in Cincinnati, Ohio. When these rumors were combined with reports that a Catholic organization in the Austrian Empire was raising funds to proselytize for Catholicism in the United States, the hackles really stood up on the necks of staunch American Protestants.

There was also a deep strand of millennial end-time belief that the Second Coming was at hand. Many American Protestants believed that they were the "chosen people" and that the Catholic church and specifically the Vatican were the "whore of Babylon" described in the biblical Book of Revelation.

One of the most prominent examples of this movement was the success of William Miller, whose nearly fatal experience in the War of 1812 had been behind a powerful conversion. Around 1830, relying upon the Book of Daniel, he calculated that the Second Coming would arrive sometime between March 1, 1843, and March 1, 1844. Tirelessly preaching that message, he eventually attracted tens of thousands of followers, and produced millions of millennial tracts distributed across America. When March 1, 1844, came and went without incident, Miller merely recalculated his "end days," determining that October 22, 1844, was the correct date. His reckoning was again mistaken, and the day became known as the "Great Disappointment." Miller died a few years later, but his ideas took hold in the movements that later became Seventh-Day Adventists and Jehovah's Witnesses.

Miller's apocalyptic vision came about the same time that Joseph Smith, Jr., who was born in Vermont in 1805, emerged with the Book of Mormon. Although poorly educated, Smith claimed to have had his first religious vision at age fourteen. In 1820, he told friends that he saw glorious beings who said all existing churches were wrong. Three years later, according to Smith, the angel Moroni revealed to him golden plates on which were written the truths of the gospel not yet revealed to man. The angel said that Smith had been chosen to prepare the world for Christ's return. Using special glasses given to him by Moroni, Smith began a three-year process of translating the plates. He sat behind a curtain and read from the plates as others recorded his words.

Smith began to attract converts immediately; but in 1838, arguments within his group and financial problems took him and many followers to Missouri. Largely because of their rumored polygamous marriages, but also because of their unconventional theology, Smith's followers were treated with suspicion in most of the locations where they settled. In 1844, Smith was arrested in a case involving polygamy and, while in jail, was killed by an angry mob. One of his followers, Brigham Young, decided to take the group farther west and settled in the Great Salt Lake Basin, which was then still Mexican territory but would later become Utah. Despite the antagonism of mainstream Christians and the federal government, the Mormons, or Church of Jesus Christ of Latter-Day Saints, prospered and grew much larger.

The flowering of these religious groups indicated a wider transformation of America's spiritual life in the early nineteenth century. Old-line Protestantism, if not under attack, was undergoing enormous changes. William Ellery Channing, born in Newport, Rhode Island, the grandson of a signer of the Declaration of Independence, was a Harvard-trained theologian who broke from traditional Calvinist orthodoxy in the early 1800s to help establish what became known as Unitarianism. Channing rejected the notion of the Trinity, believed in the essential goodness of humanity, and held that revelation might come through rational thought rather than Scripture alone. Among his followers were people like Ralph Waldo Emerson who held Transcendentalist ideas and eventually broke from the church completely. Channing's influence and success, as well as more radical departures such as Smith's Mormons and Miller's apocalyptic visions, went hand in hand with a Second Great Awakening. Like the First Great Awakening a century earlier, this one profoundly changed America.

One of the staunchest, most vocal, and most influential men to emerge from that movement was Samuel F. B. Morse. As most schoolchildren once learned, Morse is largely credited with the invention of the telegraph and the code—consisting of dots and dashes—that still bears his name. But that extraordinary combination of American ingenuity and technology would not make Morse famous—or enormously wealthy—for another few years. His famous telegraphed message—"What hath God wrought!"—was sent from Washington, D.C., to Baltimore in 1844.

Samuel was the eldest son of Jedediah Morse, a renowned Calvinist preacher who was also famous both as the author of a basic geography textbook and for his expressed belief—founded in his fundamental, Puritan-inspired Christian faith—that America would create the "largest empire that ever existed." Samuel was born in Charlestown, outside Boston, in 1791, and attended Phillips Academy in Andover, Massachusetts, and then Yale. An aspiring painter, he went to Europe to study the masters. In an incident that might seem comic if it had not had such significant repercussions, Morse was standing in a square in Rome when the pope passed by. He failed to remove his hat as the procession moved past and was struck by one of the pope's Swiss guards, who knocked his hat to the ground.

While teaching fine arts at New York University, Morse began to publish his attacks on Catholicism in the *New York Observer*, a religious newsweekly run by his brother Richard. In a series of twelve articles, Morse cataloged the abuses of Catholicism and Catholic immigrants and issued dire warnings about the fate of America. These articles were later collected and published in 1835 as a book, *The Foreign Conspiracy Against the Liberties of the United States*.

Warning of cells of Jesuit priests who were undermining American education and luring American children into Catholic schools, Morse cautioned his readers, "I exposed in my last chapter the remarkable coincidence of the tenets of Popery with the principles of despotic government, in this respect so opposite to the tenets of Protestantism; Popery, from its very nature, favoring despotism, and Protestantism, from its very nature, favoring liberty. Is it not then perfectly natural that the Austrian government should be active in supporting Catholic missions in this country? Is it not clear that the cause of Popery is the cause of despotism?"[10]

Morse's book was both controversial and influential. His biographer Kenneth Silverman writes, "Widely circulated and often extracted in the anti-Catholic press, it spurred the formation of such anti-Catholic groups as the New York Protestant Association, dedicated to exposing the inconsistency of popery with civil liberties."[11] Morse turned his Nativist views into a political career. In 1835, he formed the Native American Democratic Association ("Native American" then meant American-born whites) and became the party's chief spokesman. As a Nativist candidate for mayor in New York City, Morse was blindsided by New York's rough-and-tumble politics and finished a miserable third.

Aside from his predictions that the Catholics would bring catastrophe to American democracy, Morse had included in his warnings the titillating notion of sexual corruption in the Catholic church. He also edited an account of lechery among priests: *Confessions of a Catholic Priest*, published in 1837.[12] This was about the same time that Americans were shuddering over scandalous revelations made in another book, which was said to expose corrupt practices within the walls of a Catholic church and convent.

In January 1836, Harper Brothers published a contrived memoir called *The Awful Disclosures of Maria Monk*, written by Maria Monk. In it, she recounted a Protestant upbringing, followed by an embrace of Catholicism. But upon her arrival at the Canadian convent, Monk said, she discovered that the nuns were forced to have intercourse with lustful priests and that those who refused were murdered. She also said that the children born of these illicit unions were baptized, then strangled and thrown into a large hole in the basement.

Given complete credence by the American anti-Catholic press, the book was a sensation. Prior to the publication of Harriet Beecher Stowe's abolitionist epic, *Uncle Tom's Cabin*, in 1852, Maria Monk's salacious—by standards of the day—exposé was America's best-selling book. It was all a tissue of lies; and after a second volume appeared, Monk was discredited. She eventually turned to crime and prostitution.

But her fevered reports of abuses in convents captured the imagination of the public, apparently eager to believe the worst. Such rumors and the widespread anti-Catholic sentiment had already led to the destruction of an American convent, in Charlestown, near the Bunker Hill battlefield. The Ursuline nuns, a teaching order specializing in women's education, ran this convent. It included a boarding school for young girls, and most of the students there were Unitarians from Boston, whose well-to-do parents were more concerned with their personal than their spiritual education. But the convent aroused suspicions among the working-class farmers and tradesmen of the area. There were fears and rumors that innocent American girls were being corrupted by papist teachings.

When one of the students ran away in 1832, she told lurid stories of life in the convent. When she was followed by one of the sisters, a Protestant convert (who later returned to the convent), the girl's account of the "unholy" practices of the nuns was confirmed. The picture these two painted also confirmed the fears and suspicions of American Protestants. Tales of severe punishments, such as being forced to lick the floor or to kiss the feet of the mother superior, shocked and appalled Charlestown's Protestants.

On August 14, 1834, a mob gathered at the convent, and the mother superior unwisely warned them, "The Bishop has twenty thousand of the vilest Irishmen at his command, and you may read your riot act till your throats are sore, but you'll not quell them."

Instead of thwarting the crowd, the woman's words inflamed it. In a short time, the mob had grown, and it was soon out of control. For years, many of the locals had been hearing grotesque rumors of torture chambers and secret dungeons within the convent. The local men were eager to see these papist horrors for themselves and had also come to teach the nuns a lesson. The crowds ransacked the convent. The historian Carmine A. Prioli described the outrage that night: "As the nuns and schoolgirls were spirited away, the carnage continued. While flames roared through the convent, rioters looted and [set fire to] surrounding buildings, including the bishop's house and library. Then they broke into the mausoleum, opened the coffins and mutilated the remains of the dead."[13]

The violence inflicted on the Ursuline convent came just days after the prominent New England minister Lyman Beecher had preached the last of a series of virulently anti-Catholic sermons in Charlestown. When Beecher, the foremost Congregationalist preacher of his day, railed about the need to counteract Catholic practices, he probably did not know that some of Charlestown's men, inspired by visions of the Boston Tea Party, had already decided it was time for another Massachusetts mob to settle matters. Although Beecher's sermons have been implicated as one of the spurs of the violence which followed, the evidence suggests that the men of Charlestown had prepared their assault well before Beecher preached his three sermons. Still, Beecher did clearly voice what this Massachusetts community believed. As the Ursuline convent was consumed in flames, local fire companies called to the scene stood by and watched. Of the twelve men arrested for the violence against the convent, eleven were acquitted, and the governor later pardoned the twelfth.

A seventh-generation Puritan preacher of old Calvinist convictions, Lyman Beecher was born in New Haven in 1775. He attended Yale, where he was a prize student of Timothy Dwight, grandson of Jonathan Edwards. Beecher then began to preach on Long Island, New York. In 1806, he gained recognition for a sermon concerning Burr and Hamilton's duel. To Beecher, dueling was a vice, like alcohol, and he was vehemently opposed to both.

Beecher and his allies led a powerful reform-minded Second Awakening that was behind the creation of such groups as the American Bible Society, the American Society for the Promotion of Temperance, and the American Home Missionary Society, which together with other reform-minded associations came to be known as the "Benevolent Empire."

Within a short time, Lyman Beecher had become the most prominent Calvinist preacher of his day; and in 1832 he moved to Cincinnati—supposed site of the future Vatican—in order to found the Lane Theological Seminary. A few years later, he wrote *A Plea for the West*, a call to Americans to move west—well before Horace Greeley's more famous advice, "Go west, young man." Beecher's idea was to plant the banner of Christianity in the new territories being opened up. Part patriotic cheerleading for aggressive expansionist policies, Beecher's plea included a sharply anti-immigrant message combined with anti-Catholic vitriol. He had emerged as a leading voice of Puritan Nativism.

"A tenth part of the suffrage of the nation, thus condensed and wielded by the Catholic powers," wrote Beecher, "might decide our elections, perplex our policy, inflame and divide the nation, break the bond of our union, and throw down our free institutions. The voice of history also warns us, that no sinister influence has ever intruded itself into politics, so virulent and disastrous as that of an ambitious ecclesiastical influence."[14]

Lyman Beecher's antipapal rhetoric made clear what he thought. "Papal puppets" were threatening American freedom.

THE GREAT WAVE of anti-immigrant, anti-Catholic hatred and intolerance did not ebb as the nation entered the 1840s. The Bible Riots of 1844 were only the most deadly and visible example of this streak of secular intolerance, fear, and loathing of the foreign that had consistently appeared in American politics and culture.

During the next few years the anger would deepen as the ranks of the immigrants were swollen by events in Europe, and particularly in Ireland. "The blight came upon Ireland suddenly. As harvest time approached in 1845, the crops looked splendid," an emigrant remembered. "But one fine morning in July there was a cry around that some blight had struck the potato stalks." The leaves blackened, the tubers quickly rotted, and 'a sickly odor of decay' spread over the land. . . . They constituted a novel, unforeseen and mysterious catastrophe, producing the last major famine in European history: an gorta mór, 'the great hunger,' in Irish."[15]

In 1846–1855, more than 1 million Irish died—mostly from starvation and diseases brought on by malnutrition. Millions more emigrated—some went to England; some went to Australia and elsewhere in the British Empire; and more than 1 million came to America.

This great influx of Irish Catholics fueled Nativist suspicion and animosity. Mixed with the era's economic turmoil, the long-standing hatred of Catholicism accelerated the Nativist movement.

But there was another reality. In Andrew Jackson's Democratic Party—or at least in its northern branches—these immigrants were welcomed. In New York City they were transformed into a bloc of loyal voters that would become part of the machine politics of Tammany Hall. Similar political power brokers operated in Boston and other growing northern urban centers.

In the 1840s, after violence failed to stanch the flow of immigrants into America, the Nativist parties attempted to limit the political power of the new Americans. Regional Nativist parties began coalescing around the aims of denying the vote to noncitizens, making citizenship more difficult, and restricting political offices to native-born Americas. One of these parties, the Order of the Star-Spangled Banner, was a secret society formed in 1850. As the story goes, when its members were asked about the party they would simply say they "knew nothing." Thus, Horace Greeley of the *New York Tribune* dubbed them the "Know-Nothing Party." The name stuck.

In 1856, Anna Ella Carroll spoke for the Know-Nothings and for many other Americans in a book called *The Great American Battle; Or, the Contest between Christianity and Political Romanism*. In it, she wrote: "Roman Catholics are in the political field, fighting against American liberty; the American Party has come out to meet them in this combat! . . . I say, then, my children, the American

Party has planted its action against this political movement of the Roman Catholic Church in this dear, blood-bought land."[16]

Aftermath

Many of the newly arrived immigrants saw the army as their best chance of bed and board, and they were often recruited right off the boats. They were accustomed to military life, as conscription in Europe was commonplace. Immigrants soon made up nearly half the enlisted men in the American army and would quickly be given an opportunity to fight for their new country. America was about to embark on its first war against a neighbor. On April 4, 1846, American sentries fired the first shots in the war against Mexico—at an immigrant deserter swimming across the Rio Grande to the Mexican side of the river.

Describing the American forces, Daniel Howe writes in *What Hath God Wrought*, "The Irish alone constituted a quarter and the Germans 10 percent. The Mexicans made strong appeals to U.S. troops to switch sides, targeting immigrants and Catholics in particular. Their broadsides emphasized the injustice of the invaders' cause in the eyes of 'civilized people' and stressed what North American Catholics had in common with Mexican Catholics. Alluding to well-known riots by U.S. Protestant nativist mobs, a Mexican pamphlet asked, 'Can you fight by the side of those who put fires to your temples in Boston and Philadelphia?' "[17]

Whether any Catholics serving in the army read those pamphlets or not, many soldiers opted out of the war against America's southern neighbor. "The desertion rate in the war with Mexico, 8.3 percent, was the highest for any foreign war in United States history—twice as high as that in the Vietnam War," Howe continues. "And the prejudices of nativist officers all contributed to the desertion problem."[18]

In 1856, twelve years after the "Bible Riots," two political parties held their presidential nomination conventions in Philadelphia. The American Party, or Know-Nothings, selected the former president, Millard Fillmore, as its candidate, with Andrew Jackson Donelson, the nephew (and practically the adopted son) of Andrew Jackson, as Fillmore's running mate. The Know-Nothings had enjoyed considerable success at the local and state level, winning a couple of state houses and several seats in Congress. This party's slogan and ideology were simple: "Americans must rule America"; no immigrant should get citizenship until he had lived in America for twenty-one years; and only native-born Americans should be entitled to hold office.

A few months later, the newly formed Republican Party also met in Philadelphia. Built around a coalition of former Whigs, Free-Soilers, and abolitionists, this party was dedicated to the "free white working man." Its slogan was "Free Men, Free Speech, Free Press, Free Labor, Free Territory, and Frémont." As their first presidential candidate, the Republicans turned to one of America's most famous heroes, John Charles Frémont, known as "the pathfinder." Coming out of the gate, Frémont had two serious problems, as far as most Americans were concerned: he was for abolition, which few Americans favored, whether or not they owned slaves; and his heritage was "questionable." Opponents claimed that he wasn't American but had been born in Canada. (These opponents were nineteenth-century precursors of the "birther movement," which professed that Barack Obama was really born in Kenya.) In addition to calling the Republican candidate a drunk, a slaveholder, and a crook, the Know-Nothings claimed that Frémont was Catholic and illegitimate. They circulated pamphlets with titles such as *Frémont's Romanism Established*; *The Romish Intrigue*; *Frémont a Catholic*; and *The Authentic Account, Papist or Protestant, Which?*[19]

To parry those claims, the Republicans called upon another famous American, Henry Ward Beecher. The son of Lyman Beecher, Henry had rejected his father's Nativism and had also become one of America's most prominent abolitionists. Henry Ward Beecher would stump tirelessly for Frémont, making certain that the Republican candidate's solid Episcopal credentials were well established. He had help from his sister, Harriet Beecher Stowe, one of the most famous women in America. The author of *Uncle Tom's Cabin* and her brother, the abolitionist preacher, had eclipsed their Nativist father.

The Nativist Party's claims ultimately did not affect the outcome of the 1856 election, which was won by the Democrat, James Buchanan. But they may have led to some confusion. Frémont's supporters were so adamant in their defense of his Protestant beliefs that some voters actually thought Frémont was the Know-Nothing.

The Nativist Party gradually dissipated after the Civil War, with the emergence of the Republican Party as the dominant counterforce to the Democrats. But Nativism did not dissipate. Anti-Catholic and anti-immigrant sentiments continued as a dominant political force. By the early twentieth century, Irish and other European immigrants had gradually won greater acceptance in mainstream America; by contrast, much of the anti-immigrant anger in the post–Civil War years was turning toward the Chinese and other Asians groups. Harsh "anti-coolie" laws and other so-called "yellow laws" aimed at restricting Chinese immigration were passed in the late nineteenth century.

And although the antipathy toward Catholics eventually lost most of its violence, its force remained. After the Civil War, the Ku Klux Klan was formed; besides its fundamental racism, the Klan movement also expressed a strong aversion to foreigners, Catholics, and Jews. In the 1920s, when a Catholic, Al Smith of New York, ran against Herbert Hoover, Smith's faith was still an issue; Republican slogans claimed that he was in favor of "rum, Romanism, and ruin."

It would take more than 100 years after the first Nativist presidential candidate for Americans to elect a Roman Catholic president. In 1960, John F. Kennedy surmounted what was still a powerful suspicion of Roman Catholics. The anti-Catholic bias ran into the black community as well. During Kennedy's presidential campaign, the Reverend Martin Luther King, Sr., father of the civil rights leader, expressed reservations about Kennedy's faith, prompting Kennedy to quip, "Imagine Martin Luther King having a bigot for a father. Well, we all have fathers, don't we?"[20]

Religion has, of course, occupied a central place in American history. But its influence has not always been benign or positive. The deep divisions along some fundamental religious fault lines still run through the country. In many ways, those divisions were exposed and even created during this stormy moment in the history of a so-called Christian nation.

VI

Jessie's Journey

1811 Russian settlers land at Bodega Bay, north of San Francisco, in February and establish Fort Ross, an agricultural colony, to supply their settlements in Alaska.

1821 With the permission of Mexico, Moses Austin settles 300 American families in Texas, then Mexican territory. After his death in July, his son Stephen takes over the grant.

1829 Mexico rejects President Jackson's offer to purchase Texas.

1835 Mexico's military dictator Santa Anna includes Texas in the laws against slavery; American settlers announce plans to secede from Mexico.

1836 Following the attack by Mexican troops on the Alamo in February, Texans defeat Santa Anna, ratify a constitution, and elect Sam Houston president of the Republic of Texas in April.

1837 On his last day in office, President Andrew Jackson recognizes Texas as an independent republic.

1841 Swiss-born John Augustus Sutter purchases Fort Ross from the Russians.

1845 Texas is admitted to the Union as a slave state (the twenty-eighth state).

John C. Frémont publishes *The Report of the Exploring Expedition to the Rocky Mountains in the Year 1843 and to Oregon and Northern California in the Years 1843–44.*

1846 War is declared on Mexico in May.

In June, American settlers in California break away from Mexico and declare a Republic of California. John C. Frémont arrives on June 25 and is given command of the territory. On July 7, Commodore John Sloat arrives at Monterey and claims California for the United States.

Colonel Stephen W. Kearney arrives in Las Vegas and announces the annexation of New Mexico in August. He occupies Santa Fe and sets up a temporary government there.

1847 Promoted to general, Kearney captures Los Angeles, ending hostilities in California; Mexican forces sign a treaty with Frémont.

1848 The Treaty of Guadalupe Hidalgo ends the Mexican War. The United States receives more than 500,000 square miles of territory—including the future states of California, Nevada, Utah, and Arizona; most of New Mexico; and parts of Wyoming and Colorado, along with Texas.

Wisconsin is admitted to the Union as a free state (the thirtieth state).

General Zachary Taylor, a hero of the Mexican War, is elected the twelfth president. Van Buren's Free-Soil candidacy wins 300,000 votes and draws votes away from Lewis Cass, the Democrat.

1849 The California gold rush. By the end of the year, the population of California swells to more than 100,000.

1850 President Zachary Taylor dies of cholera in July; Millard Fillmore becomes the thirteenth president.

The Compromise of 1850: five bills aimed at settling the issue of slavery are passed.

California is admitted to the Union as a free state (the thirty-first state).

1851 Serial publication of Harriet Beecher Stowe's *Uncle Tom's Cabin* begins. The novel goes on to sell 500,000 copies in one year and more than 1 million copies after a few years.

1852 The Democrat Franklin Pierce defeats the Mexican War hero Winfield Scott, a Whig, to become the fourteenth president.

The tide leaving us, we came to anchor near the mouth of the bay, under a high and beautifully sloping hill, upon which herds of hundreds and hundreds of red deer, and the stag, with his high branching antlers, were bounding about, looking at us for a moment, and then starting off, affrighted at the noises that we made for the purpose of seeing the variety of their beautiful attitudes and motions.

—Richard Henry Dana,
Two Years before the Mast (1840)

Nothing else is talked about but the quick fortunes to be made in California.

—New York City newspaper report, 1848

"Free Soil. Free Speech. Free Men. Frémont!"

—1856 Republican campaign slogan

"Frémont. Free Niggers!"

—1856 Southern Democratic campaign slogan

Isthmus of Panama

March 1849

SHADED FROM THE blistering sun by a few palm leaves, the American woman and her seven-year-old daughter sat in a rough whaleboat on a river in Panama. With the sounds of screeching monkeys and the high-pitched screams of parrots filling the air around them, mother and child clung to each other as they were poled upriver by "crews of naked, barbarous negroes and Indians." Lying prostrate in the heat beside them was the woman's brother-in-law, who had been sent on the journey as an escort, but who now seemed deathly ill—a victim of the tropical climate, strange food, and difficult days of traveling through the rain forest.

Their whaleboat, although larger than the mahogany dugout canoes used by others in their party, moved slowly upstream, barely covering a mile per hour as the small flotilla labored against the current of the Río Chagres. From the city of Chagres, a port on

Panama's Caribbean coast, these travelers were heading south to Panama City on the Pacific coast. From there, they and thousands of other argonauts or "forty-niners" hoped to make their way to the new American paradise of California, where men just dipped pans into clear mountain streams and came away wealthy beyond their dreams. It was 1849. This was gold fever.

Like thousands of others, this sick, frightened threesome would have to survive the crossing of Panama, traveling in the footsteps of the conquistador Balboa.

In 1513, the Spanish soldier Vasco Núñez de Balboa had been the first European to cross this narrow thread of land that connects South and North America. Among the earliest in the wave of Spanish adventurers who arrived in the New World opened by Christopher Columbus's discoveries, he had helped plant the first permanent European settlement on mainland American soil on the isthmus in 1510—Santa María la Antigua del Darién.

But Balboa had heard tales of a sea and lands rich in gold. Setting out on September 1, 1513, he led a small company to cross the thick jungles and forbidding mountains of the isthmus. Fending off attacks by natives, Balboa and his men pressed through uncharted rain forests until they reached the top of a mountain range on September 25, and gazed down at the Pacific Ocean. Descending to the shore, Balboa set out in a canoe, becoming the first European known to navigate the waters of the Pacific. He claimed the waters for Spain and then named the ocean Mar del Sur (South Sea), since they had traveled south to reach it.

Returning by a different route to the Spanish bases on the Caribbean side of the isthmus, Balboa had captured enough gold and

Indian cotton goods to ensure that the king would value his adventure. He also made certain that the king's share of the spoils was returned to Spain. But like so many of the first generation of conquistadors, Balboa did not profit long from his extraordinary discovery. He became embroiled in a fierce contest with the Spanish governor of Peru over this vast potential wealth and who would control it. Arrested by troops commanded by Francisco Pizarro, who would later conquer Peru, Balboa was quickly tried for treason, despite his protest that he should be returned to Spain for a royal hearing. Along with four of his compatriots, Balboa was beheaded on January 15, 1519. Their heads were displayed on poles for several days, but the location of the five conquistadors' bodies was never discovered.

Now, in 1849, little had changed in the method of travel or the challenges of traversing the jungle of Panama. Shortly after her arrival in Chagres a few weeks earlier, before setting off for Panama City, the twenty-four-year-old mother had questioned the wisdom of her decision. "If it had not been for pure shame and unwillingness that my father should think badly of me," she later recounted, "I would have returned to New York on the steamer, as the captain begged, putting before me such a list of dangers to health and discomforts and risks of every kind to kill my courage."[1] She was still mourning the loss of a son, her second child, when she had set out on this difficult and dangerous journey.

But Jessie Benton Frémont had grown up not wanting to disappoint the first of the two most important men in her life: her father, the powerful senator Thomas Hart Benton of Missouri. And, perhaps more significantly, she did not want to fail the second: her husband, Captain John Frémont—the "great pathfinder of the West"—who she believed and hoped awaited her arrival in California.

At night, Jessie and her daughter, Lily, were given a tent, "its canvas floors and walls lit by a great fire outside," the child would later recall. Less pleasant was the food. On one occasion when they stopped in a small village called Gorgona, Jessie was appalled by the main dish, a baked monkey. She later wrote that it "looked like a little child that had been burned to death."[2] Her brother-in-law Richard's travails ended at Gorgona. As his condition seemed to be worsening, it was decided that he should return by boat to Chagres. The captain leading the expedition implored Jessie Frémont to give up the journey and join Richard. But after wavering briefly, she decided to press on.

Soon they would have to abandon even the small comforts of the shaded whaleboat to continue the journey across the forbiddingly steep, rugged interior of Panama on mules. The twenty-mile trek over the mountainous spine of the isthmus was another test of courage, since the worn mule track was barely wide enough for a single animal. The mules carried mother and daughter up the breathtaking inclines of the Central American jungle, and down again. Dead mules littered the route. Streams had to be forded—there were no bridges.

A few years after Jessie Frémont made the crossing, Emiline Day, another intrepid women heading for the goldfields of California via Panama, would describe the terrors of the mule path:

The road consists of a narrow trace, in many places only wide enough for one packed mule to pass at a time. . . . It has been traveled over by mules until they have worn a track in the earth so deep that the . . . level was far above our heads, and the track in the earth so narrow that we could touch each bank with our hands as we sat on our mules. In other places they had worn steps in solid rock six inches deep in which they stepped as regularly and with as much ease apparently as they would pass over a level road. . . . For many miles together we traveled where one false step would have precipitated us down over steep and craggy rocks to a distance several hundred feet where no human being could hope to escape alive.[3]

After two days on the mule track, Jesse and Lily Frémont found themselves in Panama City, a bustling Pacific Ocean port, hoping for passage on a steamer that would carry them to California and the expected rendezvous with John Frémont. Jessie's husband was making his way west overland on another of the expeditions of discovery and mapping that had made his name synonymous with the opening of the westward routes to California. But Jessie had no way of knowing his whereabouts or condition.

When they arrived, Panama City was bursting with thousands of men and a few women who were moving across the isthmus—begging, borrowing, stealing—to make their way to California's newly discovered goldfields.

In 1849, there were only three ways to get to California from the east coast of the United States. One was the long, guided transcontinental trek, through Indian country and the mountains and deserts of the West that had already claimed travelers like those of the Donner Party, who had perished notoriously in the winter of 1847. This was a difficult, dangerous, six-month trip, fraught with the combined dangers of attack, weather, and sickness. The second route involved sailing all the way around South America aboard a steamship or the new generation of clipper ships being built for the voyage to California and for the opening of trade with Asia. But the voyage was expensive and also took six months. The third choice was to cross the jungles of Panama. This was a shorter trip, but it was incredibly arduous. And it was the choice that Jessie Benton Frémont had made.

The journey had begun in New York, where her father saw Jessie off on the steamer *Crescent City* bound for Panama, accompanied by her brother-in-law, Richard Taylor Jacob; and her daughter, Lily. A maid had also been hired for the trip—at the last moment, when Jessie's longtime servant Harriet begged not to be forced to go on the long journey. During the sea voyage to Panama, Jessie awoke to find the newly hired maid going through her luggage. As she later recounted, "In the mirror, I saw my new maid take off her wig of plainly dressed dark hair and show herself to be years younger with light hair. She then opened my trunk, took out collars, cuffs, handkerchiefs, a little armful, and went softly out of the room. Instantly I was up . . . and locked the door." The woman proved to be a known thief, but Jessie allowed her to continue on.

In the walled Panama City, booking passage to California presented new difficulties. The crews that had sailed the ships to San Francisco often decided to stay in California, hoping to make their fortunes in the frenzy of gold fever.

Although she fortunately found lodgings with a respected Spanish woman in Panama City, Jessie began to fear that her journey would end there. The city was overcrowded with travelers desperately looking for passage to California, and she began to wonder if she would ever leave the crowded, dirty port. Many of the voyagers who had reached Panama City never made it any farther—succumbing to malnutrition, yellow fever, and any number of mysterious tropical ailments.

Her only relief came in a letter, several months old, from John. In it, he described his own difficult winter journey across the southern Colorado Rockies. Blinding snowstorms had almost kept his company from discovering the mountain pass they were seeking. After weeks of blundering through the snow-blanketed mountains, Frémont had sent men back to New Mexico to get help. But he had nearly given up hope, and one of his men had frozen to death. He sent another group back for help, and these men found the first group, freezing and starving, one of their number having died, and perhaps having provided a meal to the others. The rescue party was able to get help and return with mules and provisions, but not before one-third of Frémont's men had died in the Colorado snows.

Grateful to know that her husband was alive, though in poor condition, Jessie herself now fell ill and began to cough blood. A pair of doctors—one an American, the other local—diagnosed her condition as "brain fever." The two physicians offered opposing courses of treatment. The American doctor called for cold drinks and fresh air. But the Panamanian physician advised bleeding and a closed room. Fortunately, no leeches were available and Jessie was spared the procedure—the loss of blood might have killed her. Finally, on May 6, she heard guns and horns in the harbor. Two ships had arrived, and she and Lily would be able to continue on to California.

An American naval officer sought her out and described the chaos and debauchery that awaited her in San Francisco—gambling, prostitution, and a lack of decent housing. He also told her he had heard that her husband had injured a leg and was heading back east. Once again, she was advised to return home. But her biographer Pamela Herr describes Jessie's intense resolve: "She had promised to meet John in San Francisco, and she believed he would keep his word. She had endured so much, she wanted to see the journey through."[4] On May 18, Jessie and Lily boarded the crowded steamer *Panama*, which was carrying 400 passengers, though it had been built to accommodate only eighty.

The word "California" had been coined in 1510 by the Spanish writer Garci Rodríguez de Montalvo, in a romance about a mystical land (placed near the Garden of Eden), and its queen, Calafia:

> *I tell you that on the right-hand side of the Indies there was an island called California, which was very close to the region of the Earthly Paradise. This island was inhabited by black women, and there were no males among them at all, for their way of life was similar to that of the Amazons. This island was made up of the wildest cliffs and the sharpest precipices found anywhere in the world. The women had energetic bodies and courageous ardent hearts, and they were very strong. Their armor was made entirely out of gold which was the only metal found on the island.*[5]

From its very first use, the word "California" connoted a fantastic promise of glory and gold. The novel, too, is a fantasy, in which Queen Calafia and her women warriors go to battle wearing their golden armor, studded with the precious stones found on California Island. From its inception, then, the word was associated with conquest, indigenous people eager to convert to Christianity, and indigenous women willing to give themselves to European men.[6]

The Spanish had named California in the late 1500s, but thought it was not worth the expense of colonization. A royal order in 1606 actually prohibited exploration of the area. Not until the 1760s, when they were confronted by the threat of competition from Great Britain and Russia, did the Spanish seek to expand more aggressively into this territory north of their established provinces in Mexico. Instead of relying upon the militarism of the conquistador period that had established Spain's hold over South and Central America, the Spanish sent in the Franciscans. Under the leadership of Fra Junípero Serra, the Franciscans established a mission system, similar to the one used in settling Florida, and built the first of their missions in the future state of California at San Diego in 1769.

Serra had been born on the island of Mallorca in 1713, and was sent to a Franciscan school as a boy. Intellectually gifted, he was enrolled in the order by age sixteen and soon became a priest. Landing in Mexico in 1749, he demonstrated what would become one of his most outstanding traits—remarkable physical stamina. He insisted on walking the 200 miles from Veracruz to Mexico City despite being ill. When given the opportunity to open Alta (upper) California—the territory of the future state—he often displayed extraordinary drive and an extraordinary constitution, though he was frail and asthmatic. Serra also practiced the traditional mortifications of the flesh: heavy shirts with sharp barbs pointed inward to pierce the wearer's flesh, self-flagellation to the point of bleeding, and the application of burning candles to sear the chest. By the time of his death in 1784, he had opened nine missions.

In 1776, while men were gathering a continent away in Philadelphia to debate Thomas Jefferson's words, another group of Serra's Franciscan priests had arrived in northern California, along with soldiers and settlers, to establish Mission San Francisco de Asís, named after the order's founder, Saint Francis of Assisi. Founded on the day before Good Friday (the Day of Sorrows), the mission was better known as Nuestra Señora de Dolores (Our Lady of the Sorrows). The settlement mission and the adjoining presidio, or military compound, would be known as San Francisco. By 1823 there were twenty-one missions stretching north as far as Sonoma.

The California missions—which are still very popular tourist destinations—comprised more than simply a single church building. They were more like a medieval cathedral town. Usually built near a presidio, or frontier military fort, and a pueblo, a farming village, the mission was "at once a church, town, military fortress, school, farm, factory, and prison, typically maintained by two missionaries and a few armed soldiers," as the historian Joshua Paddison explains: "The padres attracted most of their Indian converts (called neophytes) through beads, food, and other gifts. Once baptized, however, neophytes could be held at missions against their will while padres attempted to regulate nearly every aspect of their lives, including sex, work, sleep, amusement, and religious practice."[7] Neophytes who were late for Mass were routinely lashed with a heavy leather thong. Spanish soldiers routinely raped young Indian women.

In their zeal, Serra and his followers also brought a reign of disease and terror to California that would nearly wipe out the Native population.* The number of Indians in California before and after the arrival of the Spanish can only be estimated. By 1845, their number was down to 150,000—many less than before European contact. The causes were influenza, smallpox, measles, typhoid fever, dysentery, and syphilis—along with forced labor, beatings, rape, and other forms of violence. In the first three years of Franciscan rule, one mission reported baptizing seventy-six Indians but burying 131. Figures at other missions were similar. As the scholar David E. Stannard writes, "Although the number of Indians within the Franciscan missions increased steadily from the close of those first three disastrous years until the opening decade or so of the nineteenth century, this increase was entirely attributable to the masses of native people who were being captured and force-marched into the mission compounds. Once thus confined, the Indians' annual death rate regularly exceeded their birth rate by more than two to one. . . . In short, the missions were furnaces of death."[8] Still, the methods

*Junípero Serra is still highly regarded in California history, a virtual icon of the Spanish colonial era; his statues stand in San Francisco's Golden Gate Park and in the U.S. Capitol. In 1987, Serra was beatified by Pope John Paul II, the second of three steps necessary for eventual sainthood. Many Indians and academics condemned the decision, pointing to the conditions of mission life, equivalent in some ways to life in a concentration camp; these critics also noted Serra's own justification of beating the Indians as necessary for their religious instruction and obedience.

were efficient: the mission system actually required few Spaniards or Mexicans to control the enormous territory.

The fact that priests instead of soldiers led this settlement of California made little difference to the outcome. It was a conquest, and later critics would dub Father Serra "the last conquistador."[9] In 1834, the governor of California ordered a secularization of the missions, and much of the property was turned over to the neophytes. Under the new policy, private citizens could apply for land grants that allowed them to operate ranchos, large cattle farms supplying the tallow trade. That trade brought a Harvard dropout named Richard Henry Dana aboard ship to California. He wrote of his experiences in *Two Years before the Mast* (1840), a popular best-seller that awakened Americans' interest in controlling California.

Although the Spanish had secured a toehold in California, still more dreamers came. Russians had arrived on the California coast in 1811 and negotiated the rights to establish an agricultural community to supply their settlement in Sitka, Alaska. The Russians built Fort Ross on Bodega Bay, about 100 miles north of San Francisco. The community there never flourished, and in 1841 the Russians sold the site to a Swiss-born entrepreneur, Johann Suter—whose name was Americanized as John Sutter—for a note that he never honored.

Sutter had come to California in 1839 with his own dream: to create an agricultural utopia he called New Helvetia. Becoming a Mexican citizen, he negotiated one of those land grants with the provincial governor of Monterey; it would become Sutter's Fort, a massive adobe structure with walls eighteen feet high and three feet thick, located in present-day Sacramento. Under the terms of this grant, Sutter was supposed to "prevent the robberies committed by adventurers from the

United States, to stop the invasion of savage Indians and the hunting and trapping by companies from the Columbia." Sutter's grant was intended by the California authorities to create a buffer against the mounting threats coming from American interests.

Despite this agreement over "American adventurers," Sutter's Fort became a profitable way station for the influx of Americans moving to California in the early 1840s, many of them inspired by books they carried as their "bibles": *The California Guide Book* or *Oregon and California*, writings by Jessie's husband, John Charles Frémont, repackaged for the gold rush trade.

BY THE TIME she was sequestered in Panama City in 1849, unable to complete the trip to California, Jessie Benton Frémont was arguably the most famous woman in America.* Born on May 31, 1824, near Lexington, Virginia, she was the daughter of Thomas Hart Benton, a Democratic senator from Missouri who between 1821 and 1851 was one of the most powerful men in Washington and who rose to become the high priest of America's aggressive Manifest Destiny.

Born in North Carolina in 1782, the son of a wealthy planter and landowner, Thomas Hart Benton† studied law briefly but left school at age seventeen to manage the family's estate. A few years later, he

*A few years later in 1852, that claim might rightfully belong to Harriet Beecher Stowe, after *Uncle Tom's Cabin* was published.

†Thomas Hart Benton, the senator, was the great-uncle of another notable American by that name, the painter and muralist Thomas Hart Benton (1889–1975), best known for his scenes of everyday life in the Midwest.

moved to Tennessee, attracted by the opportunity to acquire land in the opening western territories. He then returned to law studies, was admitted to the Tennessee bar, and came under the wing of Nashville's leading citizen, Andrew Jackson. During the War of 1812, Benton became Jackson's protégé and aide-de-camp, and was dispatched to Washington, D.C., to represent Jackson's interests there. During this period, Benton became involved in the notorious 1813 gun battle with Old Hickory.

Largely out of fear of Jackson, Benton relocated in 1815. For the Bentons, Nashville was no longer safe. "I am literally in hell here," wrote Thomas. "The meanest wretches under heaven to contend with: liars, affidavit-makers, and shameless cowards. All the puppies of Jackson are at work on me. . . . The scalping knife of Tecumseh is mercy compared to the affidavits of these villains. My life is in danger . . . for it is a settled plan to turn out puppy after puppy to bully me."[10]

Beating a hasty retreat from Tennessee, Benton moved to Missouri, where more new territory was being opened to American pioneers and speculators. Settling in Saint Louis, he set up a law practice while also editing the Missouri *Enquirer.* Benton's hot temper and the "code of the West" led to another duel in 1817, when a courtroom argument escalated into a challenge of honor. After an initial exchange of shots, Benton and the attorney Charles Lucas wounded each other. That first relatively harmless confrontation was followed by a exchange of insults over the duel, which led to a second challenge. This time, Lucas was shot in the heart and died. Although the story may be apocryphal, Benton supposedly later said, "I never quarrel, sir, but I do fight, sir, and when I fight, sir, a funeral follows, sir."

In 1820, when Missouri attained statehood, Thomas Hart Benton became one of its first U.S. senators. Once established in Washington, he repaired his relationship with Andrew Jackson, and later he campaigned for Old Hickory, eventually becoming the Senate leader of what had emerged as the dominant Democratic Party. Opposed to banks and paper currency, like Jackson, Benton argued for gold currency, earning himself the nickname "Old Bullion." As the chief proponent of a policy of aggressive expansion into the western territories, Benton also sponsored the Homestead Acts, which gave settlers acres of western land in exchange for the promise to cultivate it. He was also one of the chief sponsors of congressional support for Samuel F. B. Morse's telegraph.

When Jessie, the Bentons' second daughter, was born, her father left no question that he had hoped for and wanted a son. He named her in honor of his own father, Jesse Benton. Raised in Washington, D.C., the strong-willed, independent Jessie was brought up as if she were a boy. A hint of the unusual personal qualities of this young woman came in a letter from the mistress of "Miss English's Female Seminary," the preparatory school Jessie attended and derided as a "Society School": "Miss Jessie, although extremely intelligent, lacks the docility of a model student. Moreover, she has the objectionable manner of seeming to take our orders and assignments under consideration, to be accepted or disregarded by some standard of her own."[11]

There was a powerful bond between father and daughter. Benton's wife, Elizabeth, was content as a mother and homemaker. But Jessie had, from childhood, the spark of her father's wild western spirit. As a teenager, while most other girls of her age and status were thinking about cotillions and debuts, Jessie often traveled with Benton, and the most powerful senator in America introduced her to the world of political hand-to-hand combat, which he then dominated. He also shared with his daughter the dream of a nation stretching from ocean to ocean. Standing at the nexus of American political power, Jessie blossomed into an increasingly indispensable asset to the senator. Fluent in French and Spanish, she assisted with the translation of government documents and Mexican newspapers, for reports on the impending crisis over Texas. She also served as her father's secretary and took dictation at meetings. Eventually, she even wrote some of Senator Benton's speeches. When she received a proposal of marriage from President Martin Van Buren, her father was prompted to "cloister her" in the rural Georgetown Academy, notes her biographer Sally Denton.[12]

While she was home from school, at age sixteen, Jessie encountered Lieutenant John Charles Frémont, eleven years her senior. Fresh from an expedition as a surveyor mapping the territory between the Mississippi and Missouri rivers, the young officer found himself reporting to the senator most interested in expanding America's empire. The senator and the surveyor developed a mutual admiration, and Frémont must have known that Benton represented the hope of advancement. But Jessie and Frémont shared a very different feeling. When he went home that night, Frémont told a friend, the astronomer and geographer Joseph Nicollet, "I have fallen in love at first sight."[13]

Faced with her parents' disapproval of the relationship, Jessie eloped with Frémont; and a Catholic priest, the only willing clergyman they could find, secretly married them in October 1841. (That fact would later be used in the presidential campaign as evidence of Frémont's supposed Roman Catholicism.) In time, Jessie reconciled with her parents. Once his relationship with his daughter was repaired, Benton found in John C. Frémont a son-in-law who was uniquely willing and qualified to help him realize his vision of Manifest Destiny.

With Benton pulling strings to force the necessary appropriations through the Senate committees, Frémont had the complete backing of his father-in-law as he undertook command of his first expedition to mark the trails west.

Frémont set off in May 1842, leaving behind his young wife, who was pregnant. When he returned on October 29, he had crossed the Great Plains to reach the Rockies and had surveyed the South Pass, gateway to the Oregon country. A few weeks later, on November 15, 1842, Elizabeth Benton ("Lily") Frémont was born in Washington, D.C. Jessie and the infant Lily stayed with her parents, and Jessie was at home when her mother suffered a disabling stroke late in 1842. (Elizabeth Benton died twelve years later.)

With the legendary guide and trapper Kit Carson, who was thirty-two years old when they first set out, Frémont went on to command two more journeys of mapping and exploration: The second was to the Oregon Territory in 1844; and the third was to the Great Basin, the Sierras, and California in 1845. Intensely interested in the details of her husband's expeditions, Jessie became his recorder, making notes as he described his experiences. Adding human-interest touches to John's military-style reports, she edited the stories of the adventures Frémont had while exploring the West with Kit Carson. These narratives became sensationally popular and made Frémont famous as the "pathfinder of the West."

"They were inseparable and synergetic, and their teamwork turned the expedition into a wonderful adventure and a best-selling book," writes Denton. "They brought the characters alive—Kit Carson, Arapaho Indians, mountain men, fur traders and scouts—and gave drama to the landscape John had mapped and charted. On the page, the unshaven, rough-hewn explorers became heroes on a visionary quest."[14]

Having married into the family of Washington's greatest power broker, Frémont had traveled a long way from a childhood marked by poverty and scandal. His mother, Anne Beverley Whiting, was the youngest daughter of a socially prominent Virginia planter, Colonel Thomas Whiting, who had died when Anne was an infant. Her mother remarried, but Anne's stepfather soon exhausted most of her father's estate. To escape the family's financial problems, Anne moved in with an older married sister. In 1796, a marriage was arranged for the seventeen-year-old Anne with a wealthy, sixty-year-old Revolutionary War veteran. In 1810, Anne's husband hired Charles Fremon, a French immigrant who had fought with the Royalists during the French Revolution, as a tutor for his wife. Fremon and Anne were soon having an affair, and in a scandal that shocked genteel Richmond, the couple left together in July 1811.

Anne and Fremon moved first to Norfolk and later settled in Georgia, where they purchased a house. She took in boarders while Fremon taught French and dancing. On January 21, 1813, their first child, John Charles Fremon, was born. By coincidence, the Fremons were staying in a Nashville hotel in 1813, when Andrew Jackson and the Benton brothers had the shootout in which Jackson was wounded. As the infant John C. Frémont slept, his parents heard the whistling of balls and the "report of fire arms." It is not clear whether John Frémont himself added the accented *é* and *t* to his name, or if his father had done so. But it does seem certain that John Charles Frémont was illegitimate.

While preparing to lead his first expedition, Frémont sought a guide and met Kit Carson on a Missouri River steamboat in 1842. Their five-month journey, made with twenty-five men, was a complete success, and Frémont's report "touched off a wave of wagon caravans filled with hopeful emigrants" heading west.

Over the course of their next two journeys, Frémont and Carson opened up new routes to the West, and it was apparent that both Frémont and his sponsor, Senator Benton, were eyeing California as ripe for American possession. Ruled by Mexico from a distance, California was sparsely settled; and as more and more Americans moved west, it became clear that this was going to be another flash point with Mexico, which was already preparing for war with the United States over the future of Texas.

FOR THE FIRST time in its short history, the United States didn't go to war with a foreign power over independence, provocation, or global politics. This was a war fought unapologetically for territorial expansion. One young lieutenant who fought in Mexico later called the war "one of the most unjust ever waged by a stronger against a weaker nation." His name was Ulysses S. Grant.

The war with Mexico was the centerpiece of the administration of James K. Polk, perhaps the most adept of the presidents between Jackson and Lincoln. Continuing the line of Jacksonian Democrats in the White House after Tyler's abbreviated Whig administration, Polk was dubbed "Young Hickory." A slaveholding states' rights advocate from North Carolina, Polk slipped by Van Buren in the Democratic convention and was narrowly elected president in 1844. His victory was made possible because the splinter antislavery Liberty Party drew votes away from the Whig candidate Henry Clay. A swing of a few thousand votes, especially in New York State, which Polk barely carried, would have given the White House to Clay, a moderate who might have been one president capable of forestalling the breakup of the Union and the Civil War.

The 1844 presidential contest was a Manifest Destiny election. The issues were the future of the Oregon Territory, which Polk wanted to "reoccupy," and the future of Texas, which Polk wanted to "reannex," implying that Texas was part of the original Louisiana Purchase. (It wasn't.) Even before Polk's inauguration, Congress had adopted a joint resolution on his proposal to annex Texas. The congressional decision made a war with Mexico all but certain, and this suited Polk and other expansionists. In response to the American actions, Mexico severed diplomatic relations with the United States in March 1845.

Treating Texas as U.S. property, Polk sent General Zachary Taylor into the territory with about 1,500 troops in May 1845, to guard the undefined border against a Mexican invasion. After months of negotiating to buy Texas, Polk ordered Taylor to move to the bank of the Rio Grande. This so-called Army of Observation numbered some 3,500 men by January 1846—about half of the entire U.S. Army. Its mission was anything but "observation." Escalating the provocations, Polk next had Taylor cross the Rio Grande. When a U.S. soldier was found dead and some Mexicans attacked an American patrol on April 25, pro-war American papers, such as the *Union*, shrieked, "American blood has been shed on American soil." President Polk had the pretext he needed to announce to Congress, "War exists." Although Senator Benton was not an advocate of the war, a large Democratic majority in the House and Senate quickly voted to expand the army by an additional 50,000 men. America's most naked war of territorial aggression was under way.

Whig opposition to the war was muted, although as the war continued the party tried to expose it as unnecessary and argued that Polk had deceived the nation in order to provoke an unjustified war. A freshman Whig congressman from Illinois made his first speech opposing the conflict with Mexico, challenging the president to reveal the exact spot where the first blood had been spilled. Abraham Lincoln said the war was the product of "a bewildered, confounded, and miserably perplexed man." In assaulting Polk, he found few Whig friends, and he was advised to drop his vocal opposition to the war, which was very popular back home. He followed that advice but was not returned to the House. Voicing opposition to a popular war has never been a good political move in American history. Opposition should wait until the war becomes unpopular.

The war ended two weeks after Lincoln spoke.[15]

Won quickly and at relatively little expense, the Mexican War basically fulfilled the dream of Manifest Destiny.

• • •

While the war was still being fought in Mexico, a small group of Americans in Sonoma, California, were inspired to follow the lead of Texas. They had the behind-the-scenes support of America's hero, John Charles Frémont, who was in northern California on another surveying trip—an ill-disguised incursion into California. Encouraged by an administration in Washington that openly wanted to purchase California and Oregon, a motley group of American settlers calling themselves *osos*, or bears, declared an independent republic in California and raised the "Bear Flag" on June 25, 1846. According to Hampton Sides, their flag was a "slightly deformed banner fashioned from scraps of ladies' undergarments, with a grizzly bear (or 'something they called a bear' . . .) rising on its haunches, the crude image dribbled in berry juice."[16]

Moving south under orders from Commodore Robert F. Stockton, Frémont led a military expedition of 300 men to capture Santa Barbara in 1846. While they were crossing the Santa Ynez Mountains, Frémont's company nearly lost all their equipment in a rainstorm; but they made their way to the presidio and captured the town almost—though not entirely—without bloodshed.

On June 28, in an incident that recalls Andrew Jackson's treatment of the two Englishmen in Florida, Frémont captured three local men—Californios, the Spanish-speaking natives of California still under Mexican rule. Two of them were sons of the mayor of Sonoma. Kit Carson asked Frémont what to do with the men, and Frémont replied that he had no use for prisoners. Although the details were later disputed, Carson shot all three, apparently believing that this was what Frémont had ordered. Frémont later disavowed any part in the shooting. Hampton Sides writes, "Neither Carson nor Frémont mentioned anything about this little atrocity in his memoirs. It remains one of the unfathomable episodes of Carson's life."[17]

The success, with little cost, added to Frémont's now legendary stature. One newspaper, the *Intelligencer*, reported on the arrival of Frémont's band in Monterey: "They are the most daring and hardy set of fellows I ever looked upon. They are splendid marksmen, and can plant a bullet in an enemy's head with their horses at full gallop. . . . They never sleep in a house, but on the ground, with a blanket around them, their saddle for a pillow, and a rifle by their side."[18]

A few days later, Frémont led his men to Los Angeles and captured the small settlement there as well.

Soon afterward, warships of the U.S. Navy sailed into Monterey harbor and raised the American flag in place of the Bear Flag. The man credited with first raising the Stars and Stripes over California was the naval officer Joseph Warren Revere, grandson of the Revolutionary hero Paul Revere and namesake of Revere's friend and compatriot Joseph Warren, one of the first martyrs of the Revolution, killed at the Battle of Bunker Hill.

On January 16, 1847, after the Mexican-American War in California ended, Commodore Stockton appointed Frémont military governor of California. But a clash of egos and military services was about to complicate matters. Shortly after Frémont assumed the post, Brigadier General Stephen Watts Kearny of the U.S. Army, who had marched on Los Angeles after easily taking New Mexico from the Mexicans, said he had orders from the president and secretary of war to serve as governor. Kearny asked Frémont to give up the governorship. But Frémont was adamant in his refusal to submit to Kearny. Although Kearny gave him several opportunities to retract his refusal, Frémont stubbornly declined. When they returned east to Fort Leavenworth in August 1847, Kearny had Frémont arrested and bound in chains. He would stand trial for "Mutiny, disobedience, and conduct prejudicial to military discipline."

When Kearny tried to have the court-martial moved away from Washington, D.C., where Senator Benton still had powerful friends and allies, President Polk overruled him. Brought to the capital to stand trial, Frémont endured a court-martial that proved to be another media sensation, drawing the nation's attention and evoking comparisons in the press to Aaron Burr's trial in 1807. With the support of powerful Senator Benton and an overwhelmingly admiring press, Frémont was depicted as the wronged party—a legendary figure who had risked all to blaze a path west. He was the "great Pathfinder," a war hero, and dashing explorer, and a friend of the famous trapper Kit Carson, now being undermined by a man who wanted to steal Frémont's thunder for the conquest of California.

Almost unanimously, the national press sided with Frémont. When Frémont was reunited with Jessie in Saint Louis on the way to the trial, the press was there to capture the emotional scene as the young wife ran into the arms of the returning hero. One correspondent wrote that the court-martial would make Frémont "ten times more popular than ever."[19]

A battle of egos—General Kearny's, Commodore Stockton's, and Frémont's—as well as a turf war between army and navy, the trial was "a feeding frenzy for the media and political melodrama of the first order," wrote Hampton Sides. "Senator Benton roared his displeasure at the whole affair, arguing that his son-in-law had been unfairly caught in the crossfire of an interservice rivalry between a jealous army and a jealous navy."[20]

In the end, Frémont was convicted on all three counts. But the public, it seemed, shared the feelings of a columnist for the *New York Herald*: "Most men in the place of Frémont would have done precisely what he did."[21] After a visit to the White House during which Kit Carson and Jessie Frémont both pleaded John's case, President Polk approved of the decision of the court, though he accepted the court's suggestion of leniency for the American hero. Polk commuted his sentence to a dishonorable discharge.

But Frémont would have none of it. He was furious and considered his conviction an injustice and a grave dishonor. In February 1848, he wrote to Polk, threatening to resign his commission unless the president overturned his conviction. One month later, having received no reply from the president, Frémont resigned from the army. Pregnant throughout the arduous trial, Jessie had been ordered to bed. She gave birth to a son, Benton Frémont, on July 24, 1848, in Washington, D.C.

Bitter, Frémont again set his sights on the West. With Senator Benton's support, he received private backing for an expedition to map a route for a railroad to the west coast. With Jessie, Lily, and the new baby, he set off for Saint Louis. The baby died there on October 6, and both the Frémonts and Senator Benton blamed the child's death on Frémont's accuser, General Kearny.

Two weeks later, John set off for the West. The grief-stricken Jessie promised that she would meet him in California the following spring. It was this promise that brought her to Panama City in the spring of 1849.

SOMETHING ELSE HAD changed during the trial. The first rumors of the discovery of gold in California were filtering back east. Then came what seemed a divine confirmation of the popular notion that God had ordained that America should spread from coast to coast. On the morning of January 24, 1848, James Marshall, a mechanic from New Jersey who was building a sawmill for John Sutter on the American River not far from what is now Sacramento, spotted some flecks of yellow in the water. Although Marshall is always credited as the discoverer, JoAnn Levy offers a different view in *They Saw the Elephant*, claiming that Marshall and the others at the site didn't know what gold actually looked like. A woman named Jennie Wimmer who worked for Sutter was making soap, and when she saw the ore, she told Marshall, "This is gold, and I will throw it into my lye kettle . . . and if it is gold, it will be gold when it comes out." The next day, she removed the nugget from a bar of soap. And the gold rush was on.[22]

The gold rush transformed California and the United States. Tens of thousands of Americans set out for the West during the next few years, and $200 million worth of gold would be extracted from the hills of California. "Something very close to mass hysteria was the result," John Steele Gordon comments in his history of American economic power. "In 1849, about ninety thousand Americans set off for California, and as many followed in 1850. That is not far short of 1 percent of the population." He adds, "By 1860, more than two thousand banks were in operation in the United States. . . . As California gold flowed into the American economy, the money supply increased markedly. The minting of gold coins by the federal government increased, as did the issuance of bank notes based on gold reserves. Because the country had no central bank, there was no mechanism to regulate the money supply or to use monetary policy to control what Alan Greenspan would famously call 'irrational exuberance.' The result was a huge, but unsustainable boom."[23]

One man who did not profit from the boom was John Sutter. He lost money in the business ventures he attempted to set up after gold was discovered on his property, and he died poor while petitioning Congress for financial relief.

APART FROM THE profitable return on investment brought about by the gold rush, the aftermath of the Mexican War and the Oregon Treaty—signed a few years before, in 1846—produced other, less happy results. The addition of these enormous parcels of new territory made the future of slavery a bigger question; there was now that much more land to fight about. From the outset of the fighting, abolitionists such as the zealous William Lloyd Garrison of the American Anti-Slavery Society opposed the war, contending that it was waged "solely for the detestable and horrible purpose of extending and perpetuating American slavery."

The antislavery pacifist Horace Greeley also protested against the war with Mexico from its beginning; his *New York Tribune* became a leading antislavery voice. Another ornery gadfly went to jail in Massachusetts for his refusal to pay poll taxes that supported a war he feared would spread slavery. Henry David Thoreau spent only a single night in jail—an aunt paid his fine—but his lecture "Resistance to Civil Government" (later titled "Civil Disobedience") was published in 1849 in the book *A Week on the Concord and Merrimack Rivers*.

Perhaps the most horrible effect of the war with Mexico was the practical battle experience it provided for a corps of young American officers who fought as comrades in Mexico, only to face each other in battle fifteen years later in the Civil War. Among the many young West Pointers who fought in Mexico were two lieutenants—P. T. Beauregard and George McClellan—who served on General Scott's staff. Beauregard would lead the attack on Fort Sumter that began the Civil War. McClellan later commanded the armies of the North. Two comrades at the battle of Churubusco were lieutenants James Longstreet and Winfield Scott Hancock. They would face each other at Gettysburg in 1863. A young captain, Robert E. Lee, demonstrated his considerable military ability as one of Scott's engineers. A few years later, Scott would urge Lincoln to give Lee command of the Union armies, but Lee chose instead to remain loyal to his home, Virginia. When Lee and Ulysses S. Grant met years later at Appomattox Court House, Grant would remind Lee that they had once encountered each other as comrades in Mexico.

Aftermath

When Mexico lost the war, Spain's holdings in what became the United States of America were gone. An empire that once spanned much of the continent had disappeared. Florida, Texas, New Orleans, the Southwest, and California had all been grabbed from Spanish control. For centuries, since the establishment of Saint Augustine in 1565, the Spanish had struggled to strengthen their hold on their expansive North American territories, stretching from Florida across the continent to northern California. That struggle put Spain at odds first with the British and later with the Americans. But the once mighty Spanish, rulers of the first empire on which the sun never set, had crumbled. A shadow of its former glory, Spain ceded control of Florida to the United States in 1821, the same year it surrendered Mexico—its former source of amazing wealth and power—to independence. After being an American presence for nearly three centuries, Spain had ultimately become the loser.

In California, Jessie and John Charles Frémont proved to be among the winners. He had purchased a seventy-square-mile ranch called Las Mariposas in the Sierra foothills on a hunch, for $3,000. It proved to be a rich goldfield, and the Frémonts were soon wealthy beyond their wildest dreams. With income from their gold mines, they established a home in Monterey and later settled into San Francisco society. Flush with their gold, they purchased a prefabricated house from China with smooth, wooden interlocking pieces that were put together like a puzzle. Her father's daughter, Jessie was soon immersed in city politics. Of the experience, she later wrote, "I had done so many things that I had never done before that a new sense of power had come to me." Famous, wealthy, the son-in-law of a powerful man, John had his pick of offices. He served from September 9, 1850, to March 3, 1851, as one of California's first senators.

Then in 1856, Frémont emerged as a national political force on an antislavery platform. With America edging closer to the Civil War, Frémont's antislavery position was instrumental in his being chosen as the first Republican candidate for president. Jessie played an extremely active role in the campaign, rallying support for her husband. One campaign slogan read, "Frémont and Jessie too." However, her father, a lifelong Democrat and a proponent of slavery, refused to endorse Frémont's bid for the presidency.

While Frémont collected many northern states, not a single southern state went Republican. Nor did he carry his adopted home state, California. Although Frémont did surpass the candidate of the American, or Know-Nothing, Party, Millard Fillmore, the election went to the Democrat, James Buchanan, who is now generally considered one of the two or three worst American presidents.

Once the war came in 1861, Frémont would go on to serve—but once again, not without controversy. While commanding Union troops in Missouri, he announced an emancipation plan that predated Abraham Lincoln's, earning the president's wrath. Frémont was sacked. In 1864, he was put at the head of a dissident Republican ticket that wanted a more radical approach to abolition. Eventually, Frémont was persuaded to quit the race, and Lincoln was reelected.

Lincoln would later say of the general who had challenged him, "I have great respect for General Frémont. But the fact is that the pioneer in any movement is not generally the best man to carry that movement to a successful issue. It was so in olden times, was it not?

Moses began the emancipation of the Jews, but did not take Israel into the promised land, after all."

After the war, the Frémonts moved to New York. But when a postwar panic struck the U.S. economy in 1873, Frémont, who had invested heavily in railroad stock, lost everything and was forced to declare bankruptcy. To support the family, Jessie began a career as a writer and produced a string of successful books, including *A Year of American Travel: Narrative of Personal Experience* (1878), an account of her harrowing journey to California in 1849.

Near poverty, Frémont was appointed governor of the Territory of Arizona from 1878 to 1881. After being granted a small pension, the seventy-seven-year-old Frémont died in 1890 in a New York hotel.

After her husband's death, Jessie Benton Frémont was struggling financially as well. When a newspaper called attention to her poverty, Congress granted her a pension of $2,000 a year. In 1891, she moved into a home at the corner of 28th and Hoover Streets in Los Angeles, which was provided for her by a committee of women of the city.

Once, when asked about her husband, she told an interviewer, "Time will vindicate General Frémont. I am past sixty-six years old. I may not live to see his enemies sitting in homage at the unveiling of his statue, as in the case of my father, but John C. Frémont's name can never be erased from the most colorful chapters of American history. From the ashes of his campfires, cities have sprung."

Jessie Benton Frémont died at age seventy-eight on December 27, 1902.

She had blazed a path all the way to the twentieth century.

Acknowledgments

As I wrote at the conclusion of *America's Hidden History*, leaving the safety of familiar terrain and venturing into the unknown can be a daunting business. For me, leaving the comfortable landscape of the "Don't Know Much About" series to write about American history in a different style and format was both exhilarating and scary. But as the stories in this book demonstrate, setting out for new territory has been an essential ingredient of the American experience for centuries. And as many of these stories also prove, the results can be tragic.

With that in mind, I am again grateful for the assistance of many friends and colleagues who provided guidance and companionship on the difficult journey that is a new book. It would be impossible for me to make these journeys without the encouragement and support of a great many people who have helped me out at every step along the way.

That large group of people begins with some strangers: the many librarians at institutions large and small—from the New York Public Library to my little Dorset public library—who are committed to knowledge, learning, and the book. I also honor the people who work passionately at historic sights around the country. They provided much valuable insight and information.

David Black, my dear friend and literary agent, and his excellent team at the David Black Agency have also provided stalwart assistance over the years. I am very happy and grateful to have Dave Larabell, Leigh Ann Eliseo, Susan Raihofer, Gary Morris, Joy Tutela, and Antonella Iannarino behind me.

In the time that the "Don't Know Much About" series has been published at HarperCollins, I have also been very lucky to have the support of a dedicated publishing group behind me as well. For their continued support, I heartily thank Carrie Kania, Diane Burrowes, Leslie Cohen, Elizabeth Harper, Jen Hart, Hope Inelli, Carl Lennertz, Nicole Reardon, Michael Signorelli, and Virginia Stanley. I am also indebted to my publicist, Laura Reynolds.

It has also been my privilege to meet and work with an editor of great skill, intelligence, and enthusiasm. Elisabeth Dyssegaard once again played a crucial role in shaping and recasting this work. I greatly value her judgment and wish her well. I am also grateful to the other members of the HarperCollins team: Kathryn Whitenight, Matthew Inman, John Jusino, Susan Gamer, Mary Speaker, Karen Lumley, Richard Ljoenes, and Ben Loehnen.

My children, Colin Davis and Jenny Davis, have always provided me with joy and inspiration. I treasure their wonderful spirits and the intellectual challenges they bring to our dinner table.

And, finally, this "Hidden History" venture really started many years ago, when my wife, Joann, said to me, "You love American history. Why don't you write about it?" That's how it all began. And in this and every other journey we have shared, she has been "constant as a northern star." No pioneer could ask for a better guide and companion.

NOTES

INTRODUCTION: "THE DREAM OF OUR FOUNDERS"

1. Cited in Schlesinger, *1,000 Days.*
2. Burstein, *The Passions of Andrew Jackson*, p. xix.

I. BURR'S TRIAL

1. Isenberg, *Fallen Founder*, p. 321
2. Ibid., p. 321.
3. Chernow, *Alexander Hamilton*, pp. 303–304.
4. Gordon, *An Empire of Wealth*, p. 75.
5. Wheelan, *Jefferson's Vendetta*, p. 5.
6. Lucas, *The Aaron Burr Treason Trial*, p. 10.
7. Wheelan, p. 1.
8. Isaacson, *Benjamin Franklin: An American Life*, p. 110.
9. Witham, *A City Upon a Hill*, p. 57.
10. Wheelan, p. 27.
11. Herman, *How the Scots Invented the Modern World*, p. 243.

12. Fleming, *Liberty*, p. 154.
13. Wheelan, p. 29.
14. Isenberg, p. 36.
15. Ibid., p. 72.
16. Wheelan, p. 34.
17. Ibid., p. 32.
18. Isenberg, p. 170.
19. Gordon, p. 117.
20. Isenberg, pp. 218–219.
21. Ibid., p. 232.
22. Chernow, p. 704.
23. Isenberg, pp. 265–266.
24. Fleming, *Duel*, p. 345.
25. Cited in Isenberg, p. 277.
26. Smith, *John Marshall: Definer of a Nation*, p. 352.
27. Ibid., p. 354.
28. Burstein, The *Passions of Andrew Jackson*, p. 73.
29. Collier and Collier, *Decision in Philadelphia*, p. 231.
30. Ibid., pp. 157–158.
31. Wheelan, pp. 285–286.
32. James Parton, *The Life and Times of Aaron Burr*, Vol. 2, cited in Lucas, p. 100.

II. Weatherford's War

1. O'Brien, *In Bitterness and in Tears*, p. x.
2. Ibid., p. xi.
3. Ibid., pp. xii–xiii.

4. Ibid., p. xiii.
5. Cited in Borneman, *1812: The War That Forged a Nation*, p. 146.
6. Meacham, *American Lion*, p. xxii.
7. Remini, *Andrew Jackson*, pp. 4–5.
8. Ibid., p. 9.
9. Richter, *Facing East from Indian Country*, p. 178.
10. Ibid., p. 226.
11. Utley and Washburn, *Indian Wars*, p.117.
12. Calloway, *The Shawnees and the War for America*, pp. 138–139.
13. Utley and Washburn, p. 126.
14. Borneman, *1812: The War That Forged a Nation*, p. 1.
15. An excellent summary of this case is available in Gordon-Reed, *Thomas Jefferson and Sally Hemings: An American Controversy*, pp. 59ff.
16. Cited in Boller, *Presidential Campaigns*, p. 20.
17. Cited in O'Brien, p. 63.
18. Cited in Richter, p. 228.
19. Remini, *Andrew Jackson*, p. 57.
20. Reynolds, *Waking Giant*, pp. 2–3.
21. Cited in O'Brien, p. 150.
22. Cited in Borneman, p. 152.
23. Anderson and Cayton, *The Dominion of War*, p. 234.
24. Ibid., p. 236.
25. Burstein, *The Passions of Andrew Jackson*, pp. 131–133.
26. O'Brien, p. 240.
27. Burstein, p. 133.

III. Madison's Mutiny

1. Berlin, *Many Thousands Gone*, p. 362.
2. Rediker, *The Slave Ship*, p. 292.
3. Hendrick and Hendrick, *The Creole Mutiny*, pp. 11–12.
4. Rediker, p. 292.
5. Northup, *Twelve Years a Slave: Narrative of Solomon Northup, a Citizen of New York, Kidnapped in Washington City in 1841, and Rescued in 1853.* Electronic Edition, University of North Carolina Press, 1997.
6. Jones, *Mutiny on the Amistad*, p. 17.
7. Irons, *A People's History of the Supreme Court*, p. 151.
8. Lepore, *New York Burning*, p. xii.
9. Hochschild, *Bury the Chains*, p. 269
10. Ibid., p. 279. This is also my source for Leclerc's warning, below: "You will have to exterminate . . ."
11. Cited in Higginson, *Black Rebellion: Five Slave Revolts*, p. 72.
12. Egerton, *Gabriel's Rebellion*, p. 101.
13. Berlin, p. 362.
14. Hendrick and Hendrick, p. 110.
15. Howe, *What Hath God Wrought*, pp. 673–674.
16. Mayer, *All on Fire*, p. 316.

IV. Dade's Promise

1. Cited in O'Brien, *In Bitterness and in Tears*, p. 239.
2. Laumer, *Dade's Last Command,* pp. 177ff.
3. Ibid.
4. Mahon, *History of the Second Seminole War*, p. 105.
5. Knetsch, *Florida's Seminole Wars*, p. 72.

6. Ibid., p. 82.
7. Laumer, p. 15.
8. Meacham, *American Lion*, p. 95.
9. Ibid.
10. Cited in Meltzer, *Hunted Like a Wolf*, pp. 818–820.
11. Oates, *The Fires of Jubilee*, p. 15.
12. Reynolds, *John Brown, Abolitionist*, p. 52.
13. Oates, p. 25.
14. Ibid., p. 122.
15. Stampp, *The Peculiar Institution*, p. 134.
16. Reynolds, p. 241.
17. Price, *Maroon Societies*, p. 15.
18. Cited in Bennet, *The Shaping of Black America*, p. 106.
19. Utley and Washburn, *Indian Wars*, p. 131.
20. Ibid., pp. 130–131.
21. Cited in O'Brien, p. 239.
22. Cited in Meltzer, p. 137.
23. Mahon, p. 325.
24. Meacham, p. 92.

V. Morse's Code

1. The Project Gutenberg eBook of *Awful Disclosures*, by Maria Monk. Copyright laws are changing all over the world. Be sure to check the copyright laws for your country before downloading or redistributing this or any other Project Gutenberg eBook.
2. Clark, *The Irish in Philadelphia*, p. 17.
3. Peter Quinn, "Immigration's Dark History," *America Magazine*, February 18, 1995.

4. "The Riots," *Pennsylvania Freeman*, July 18, 1844.
5. Howe, *What Hath God Wrought*, p. 505.
6. *Life and Times in Colonial Philadelphia,* cited in Brands, *The First American*, pp. 216–217.
7. Brookhiser, *Gentleman Revolutionary: Gouverneur Morris—The Rake Who Wrote the Constitution*, p. 32.
8. McCullough, *John Adams*, p. 505.
9. Gaustad and Schmidt, *The Religious History of America*, p. 170.
10. Morse, *Foreign Conspiracy against the Liberties of the United States.*
11. Silverman, *Lightning Man*, p. 136.
12. Ibid., p. 138.
13. Carmine A. Prioli, "The Ursuline Outrage," *American Heritage,* February/March 1982.
14. Beecher, *A Plea for the West.*
15. Howe, p. 822.
16. Carroll, *The Great American Battle*, p. 129.
17. Howe, pp. 751–752.
18. Ibid.
19. Boller, *Presidential Campaigns*, p. 97.
20. Schlesinger, *A Thousand Days*, p. 74.

VI. JESSIE'S JOURNEY

1. Frémont, *A Year of American Travel*, p. 26.
2. Ibid.
3. Cited in Levy, *They Saw the Elephant*, p. 42.
4. Herr, *Jessie Benton Frémont*, p. 194.
5. Excerpted in Beebe and Senkewicz, *Lands of Promise and Despair,* p. 11.

6. Ibid., p. 10.
7. Paddison, *A World Transformed*, p. xiii.
8. Stannard, *American Holocaust*, pp. 136–137.
9. Ibid., p. 139.
10. Cited in Brands, *Andrew Jackson: His Life and Times*, p. 190.
11. Denton, *Passion and Principle*, p. x.
12. Ibid.
13. Ibid., p. xiii.
14. Ibid., p. 85.
15. Lincoln's speech cited in Oates, *With Malice toward None*, pp. 86–87.
16. Sides, *Blood and Thunder*, p. 117.
17. Ibid., p. 120.
18. Cited in Herr, *Jessie Benton Frémont*, pp. 145–146.
19. Denton, p. 143.
20. Sides, p. 255.
21. Cited in Herr, p. 173.
22. Cited in Levy, pp. xx–xxi.
23. Gordon, *An Empire of Wealth*, pp. 183–184.

Bibliography

Abernethy, Thomas Perkins. *The Burr Conspiracy.* New York: Oxford University Press, 1954.

Adams, Henry. *History of the United States of America during the Administrations of Thomas Jefferson: 1801–1809.* New York: Library of America, 1986.

———. *History of the United States of America during the Administrations of James Madison: 1809–1817.* New York: Library of America, 1986.

Amar, Akhil Reed. *America's Constitution: A Biography.* New York: Random House, 2005.

Ambrose, Stephen E. *Undaunted Courage: Meriwether Lewis, Thomas Jefferson, and the Opening of the American West.* New York: Simon and Schuster, 1996.

Anderson, Fred, and Andrew Cayton. *The Dominion of War: Empire and Liberty in North America, 1500–2000.* New York: Viking Penguin, 2005.

Applegate, Debby. *The Most Famous Man in America: The Biography of Henry Ward Beecher.* New York: Doubleday, 2006.

Beebe, Rose Marie, and Robert M. Senkewicz, eds. *Lands of Promise and Despair: Chronicles of Early California, 1535–1846.* Berkeley, Calif.: Heyday, 2001.

———. *Testimonios: Early California through the Eyes of Women, 1815–1848.* Berkeley, Calif.: Heyday, 2006.

Beecher, Lyman. *A Plea for the West.* Carlisle, Mass.: Applewood, 2009.

Bennet, Lerone, Jr. *The Shaping of Black America.* New York: Penguin, 1993.

Berlin, Ira. *Many Thousands Gone: The First Two Centuries of Slavery in North America.* Cambridge, Mass.: Harvard University Press, 1998.

Blight, David W. *A Slave No More: Two Men Who Escaped to Freedom, Including Their Own Narratives of Emancipation.* Boston, Mass.: Houghton Mifflin, 2007.

Boller, Paul F., Jr. *Presidential Campaigns.* New York: Oxford University Press, 1985.

———. *Presidential Wives: An Anecdotal History.* New York: Oxford University Press, 1998.

Boorstin, Daniel J., ed. *An American Primer.* Chicago, Ill.: University of Chicago Press, 1966.

Borneman, Walter R. *1812: The War That Forged a Nation.* New York: HarperCollins, 2004.

Brands, H. W. *The Age of Gold: The California Gold Rush and the New American Dream.* New York: Anchor, 2004.

———. *Andrew Jackson: His Life and Times.* New York: Anchor, 2006.

———. *The First American: The Life and Times of Benjamin Franklin.* New York: Doubleday, 2000.

Brookhiser. Richard. *Founding Father: Rediscovering George Washington.* New York: Free Press, 1996.

———. *Gentleman Revolutionary: Gouverneur Morris, the Rake Who Wrote the Constitution*. New York: Free Press, 2003.

Brown, Dee. *Bury My Heart at Wounded Knee: An Indian History of the American West.* New York: Henry Holt, 1970.

Burrowes, Edwin G., and Mike Wallace. *Gotham: A History of New York City to 1898.* New York: Oxford University Press, 1999.

Burstein, Andrew. *America's Jubilee: How in 1826 a Generation Remembered Fifty Years of Independence.* New York: Knopf, 2001.

———. *The Original Knickerbocker: The Life of Washington Irving.* New York. Basic Books, 2007.

———. *The Passions of Andrew Jackson.* New York: Vintage Books, 2004.

Calloway, Colin G. *First Peoples: A Documentary Survey of American Indian History.* Boston, Mass.: Bedford/St. Martin's, 2004.

———. *The Shawnees and the War for America*. Penguin Library of American Indian History. New York: Penguin, 2007.

Carroll, Andrew, ed. *Letters of a Nation.* New York: Kodansha, 1997.

Carroll, Anna Ella. *The Great American Battle: or, The Contest between Christianity and Political Romanism.* New York: Miller, Orton, and Mulligan, 1856.

Chaffin, Tom. *Pathfinder: John Charles Frémont and the Course of American Empire.* New York: Hill and Wang, 2002.

Chernow, Ron. *Alexander Hamilton.* New York: Penguin, 2004.

Christian, Charles M. *Black Saga: The African American Experience: A Chronology.* Washington, D.C.: Civitas, 1999.

Clark, Dennis. *The Irish in Philadelphia: Ten Generations of Urban Experience.* Philadelphia, Pa.: Temple University Press, 1973.

Collier, Christopher, and James Lincoln Collier. *Decision in Philadelphia: The Constitutional Convention of 1787.* New York: Random House, 1986.

Davis, David Brion. *Inhuman Bondage: The Rise and Fall of Slavery in the New World.* New York: Oxford University Press, 2006.

Davis, William C. *Three Roads to the Alamo: David Crockett, James Bowie, and William Barret Travis.* New York: HarperCollins, 1998.

Dawdy, Shannon Lee. *Building the Devil's Empire: French Colonial New Orleans.* Chicago, Ill.: University of Chicago Press, 2008.

Denton, Sally. *Passion and Principle: John and Jessie Frémont, the Couple Whose Power, Politics, and Love Shaped Nineteenth-Century America.* New York: Bloomsbury, 2007.

De Voto, Bernard. *The Year of Decision: 1846.* New York: St. Martin's, 2000.

Douglass, Frederick. *Autobiographies (Narrative of the Life of Frederick Douglass, an American Slave; My Bondage and My Freedom; Life and Times of Frederick Douglass).* New York: Library of America, 1994. (A collection of all three of Douglass's memoirs with additional editorial and historical information, annotated by Henry Louis Gates. There are many other editions of these books available individually.)

Drimmer, Frederick, ed. *Captured by the Indians: 15 Firsthand Accounts, 1750–1870.* New York: Dover, 1961.

Du Bois, W. E. B. *The Souls of Black Folk.* New York: Library of America, 1990.

Egerton, Douglas. *Gabriel's Rebellion: The Virginia Slave Rebellion of 1800 and 1802.* Chapel Hill: University of North Carolina Press, 1993.

Ellis, Joseph J. *American Sphinx: The Character of Thomas Jefferson.* New York: Knopf, 1997.

———. *Founding Brothers: The Revolutionary Generation.* New York: Knopf, 2001.

Emerson, Ralph Waldo. *Essays: First and Second Series*. New York: Library of America, 1990.

Fehrenbach, T. R. *Lone Star: A History of Texas and the Texans*. Updated ed. Cambridge, Mass.: Da Capo, 2000.

Fitch, Raymond E., ed. *Breaking with Burr: Harman Blennerhasset's Journal, 1807.* Athens: Ohio University Press, 1998.

Fleming, Thomas. *Duel: Alexander Hamilton, Aaron Burr, and the Future of America*. New York: Basic Books, 1999.

———. *Liberty: The American Revolution*. New York: Viking, 1997.

Franklin, John Hope, and Alfred A. Moss, Jr. *From Slavery to Freedom: A History of African Americans*. 8th ed. New York: Knopf, 2009.

Frémont, Jessie Benton. *A Year of American Travel*. New York: Harper, 1878.

Gaustad, Edwin, and Leigh Schmidt. *The Religious History of America: The Heart of the American Story from Colonial Times to Today*. Rev. ed. New York: HarperCollins, 2002.

Genovese, Eugene D. *Roll, Jordan, Roll: The World the Slaves Made.* New York: Random House, 1974.

Goerke, Betty. *Chief Marin: Leader, Rebel, and Legend: A History of Marin County's Namesake and His People*. Berkeley, Calif.: Heyday, 2007.

Gordon, John Steele. *An Empire of Wealth: The Epic History of American Economic Power*. New York: HarperCollins, 2004.

Gordon-Reed, Annette. *Thomas Jefferson and Sally Hemings: An American Controversy.* Charlottesville: University Press of Virginia, 1997.

Gould, Lewis L. *Grand Old Party: A History of the Republicans*. New York: Random House, 2003.

Gunn, Giles, ed. *Early American Writing.* New York: Penguin, 1994.

Hamilton, Alexander. *Writings*. New York: Library of America, 2001.

Heidler, David S., and Jeanne T. Heidler. *Old Hickory's War: Andrew Jackson and the Quest for Empire*. Baton Rouge: Louisiana State University Press, 2003.

Hendrick, George, and Willene Hendrick. *The* Creole *Mutiny: A Tale of Revolt Aboard a Slave Ship*. Chicago, Ill.: Ivan R. Dee, 2003.

Herman, Arthur. *How the Scots Invented the Modern World: The True Story of How Western Europe's Poorest Nation Created Our World and Everything in It*. New York: Three Rivers, 2001.

Herr, Pamela. *Jessie Benton Frémont: A Biography*. Norman: University of Oklahoma Press, 1987.

Higginson, Thomas Wentworth. *Black Rebellion: Five Slave Revolts*. New York: Da Capo, 1998.

Hirshson, Stanley P. *The White Tecumseh: A Biography of William T. Sherman*. New York: Wiley, 1997.

Holliday, J. S. *The World Rushed In: The California Gold Rush Experience*. Norman: University of Oklahoma Press, 1981.

Hochschild, Adam. *Bury the Chains: Prophets and Rebels in the Fight to Free an Empire's Slaves*. Boston, Mass.: Houghton Mifflin, 2005.

Howe, Daniel Walker. *What Hath God Wrought: The Transformation of America, 1815–1848*. New York: Oxford University Press, 2007.

Irons, Peter. *A People's History of the Supreme Court: The Men and Women Whose Cases Have Shaped Our Constitution*. New York: Viking Penguin, 1999.

Isaacson, Walter. *Benjamin Franklin: An American Life*. New York: Simon and Schuster, 2003.

Isenberg, Nancy. *Fallen Founder: The Life of Aaron Burr*. New York: Viking, 2007.

Jacoby, Susan. *Freethinkers: A History of American Secularism*. New York: Henry Holt, 2004.

James, Marquis. *The Raven: A Biography of Sam Houston*. Austin: University of Texas Press, 1929.

Jefferson, Thomas. *Jefferson: Public and Private Papers*. New York: Library of America, 1990.

Jones, Howard. *Mutiny on the* Amistad. New York: Oxford University Press, 1987.

Kennedy, Roger G. *Burr, Hamilton, and Jefferson: A Study in Character*. New York: Oxford University Press, 2000.

Kluger, Richard. *Seizing Destiny: How America Grew from Sea to Shining Sea*. New York: Knopf, 2007.

Kly, Y. N., ed. *The Invisible War: The African American Anti-Slavery Resistance from the Stono Rebellion through the Seminole Wars*. Atlanta, Ga.: Clarity Press, 2006.

Knetsch. Joe. *Florida's Seminole Wars: 1817–1858*. Charleston, S.C.: Arcadia, 2003.

Kolchin, Peter. *American Slavery: 1619–1877*. New York: Hill and Wang, 1993.

Laumer, Frank. *Dade's Last Command*. Gainesville: University Press of Florida, 1995.

Langguth, A. J. *Union 1812: The Americans Who Fought the Second War of Independence*. New York: Simon and Schuster, 2006.

Lepore, Jill. *New York Burning: Liberty, Slavery, and Conspiracy in Eighteenth-Century Manhattan*. New York: Knopf, 2005.

Levy, JoAnn. *They Saw the Elephant: Women in the California Gold Rush*. Norman: University of Oklahoma Press, 1990.

Lockhart, James, and Stuart B. Schwartz. *Early Latin America: A History of Colonial Spanish America and Brazil*. New York: Cambridge University Press, 1999.

Lucas, Eileen. *The Aaron Burr Treason Trial: A Headline Court Case*. Berkeley Heights, N.J.: Enslow, 2003.

McCaleb, Walter Flavius. *The Aaron Burr Conspiracy: A History Largely from Original and Hitherto Unused Sources*. Honolulu, Hawaii: University Press of the Pacific, 2006. (Reprinted from the 1903 edition.)

McCullough, David. *John Adams*. New York: Simon and Schuster, 2001.

Madison, James. *Writings*. New York: Library of America, 1999.

Mahon, John K. *History of the Second Seminole War: 1835–1842*. Rev. ed. Gainesville: University Press of Florida, 1985.

Mayer, Henry. *All on Fire: William Lloyd Garrison and the Abolition of Slavery*. New York: Norton, 1998.

Meacham, Jon. *American Lion: Andrew Jackson in the White House*. New York: Random House, 2008.

Melton, Buckner F., Jr. *Aaron Burr: Conspiracy to Treason*. New York: Wiley, 2002.

Meltzer, Milton. *Hunted Like a Wolf: The Story of the Seminole War*. Sarasota, Fla.: Pineapple, 2004.

Michno, Gregory, and Susan Michno. *A Fate Worse Than Death: Indian Captivities in the West*. Caldwell, Ida.: Caxton, 2007.

Miller, Nathan. *Star-Spangled Men: America's Ten Worst Presidents*. New York: Scribner, 1998.

Mihm, Stephen. *A Nation of Counterfeiters: Capitalists, Con Men, and the Making of the United States*. Cambridge, Mass.: Harvard University Press, 2007.

Northup, Solomon. *Twelve Years a Slave*. Radford, Va.: Wilder, 2008.

Oates, Stephen B. *The Fires of Jubilee: Nat Turner's Fierce Rebellion*. New York: Harper and Row, 1975.

———. *With Malice Toward None: The Life of Abraham Lincoln*. New York: Harper and Row, 1977.

O'Brien, Sean Michael. *In Bitterness and in Tears: Andrew Jackson's Destruction of the Creeks and Seminoles*. Guilford, Conn.: Lyons/Globe Pequot, 2003.

Paddison, Joshua, ed. *A World Transformed: Firsthand Accounts of California Before the Gold Rush*. Berkeley, Calif.: Heyday, 1995.

Page, Jake. *In the Hands of the Great Spirit: The 20,000 Year History of American Indians*. New York: Free Press, 2003.

Price, Richard, ed. *Maroon Societies: Rebel Slave Communities in the Americas*. 3rd ed. Baltimore, Md.: Johns Hopkins University Press, 1996.

Quarles, Benjamin. *The Negro in the Making of America*. New York: Macmillan, 1987.

Rediker, Marcus. *The Slave Ship: A Human History*. New York: Viking, 2007.

Remini, Robert V. *Andrew Jackson*. New York: HarperPerennial, 1999.

———. *Andrew Jackson and His Indian Wars*. New York: Viking, 2001.

———. *The House: The History of the House of Representatives*. New York: Smithsonian Books, 2006.

———. *The Life of Andrew Jackson*. New York: Harper and Row, 1988. (Condensation of the author's three-volume biography of Jackson.)

Reynolds, David S. *John Brown, Abolitionist: The Man Who Killed Slavery, Sparked the Civil War, and Seeded Civil Rights*. New York: Knopf, 2005.

———. *Waking Giant: America in the Age of Jackson*. New York: HarperCollins, 2008.

———. *Walt Whitman's America: A Cultural Biography*. New York: Vintage, 1996.

Richards, Rand. *Historic San Francisco: A Concise History and Guide*. San Francisco, Calif.: Heritage House, 2007.

Richter, Daniel K. *Facing East from Indian Country*. Cambridge, Mass: Harvard University Press, 2001.

Roberts, Cokie. *Founding Mothers: The Women Who Raised Our Nation*. New York: HarperCollins, 2004.

Rodriguez, Junius P., ed. *Encyclopedia of Slave Resistance and Rebellion*. Westport, Conn.: Greenwood, 2006.

Rogow, Arnold A. *A Fatal Friendship: Alexander Hamilton and Aaron Burr*. New York: Hill and Wang, 1998.

Schlesinger, Arthur M., Jr. *A Thousand Days: John F. Kennedy in the White House*. Boston, Mass.: Houghton Mifflin, 1965.

Sides, Hampton. *Blood and Thunder: The Epic Story of Kit Carson and the Conquest of the American West*. New York: Doubleday, 2006.

Silverman, Kenneth. *Lightning Man: The Accursed Life of Samuel F. B. Morse*. Cambridge, Mass.: Da Capo, 2003.

Smith, Jean Edward. *John Marshall: Definer of a Nation*. New York: Henry Holt, 1996.

Stampp, Kenneth M. *The Peculiar Institution: Slavery in the Ante-Bellum South*. New York: Vintage, 1984.

Stannard, David E. *American Holocaust: The Conquest of the New World*. New York: Oxford University Press, 1992.

Stone, Ilene, and Suzanna M. Grenz. *Jessie Benton Frémont: Missouri's Trailblazer*. Columbia: University of Missouri Press, 2005.

Tocqueville, Alexis de. *Democracy in America*. Trans. George Lawrence. New York: Perennial, 1988.

Utley, Robert M., and Wilcomb E. Washburn. *Indian Wars.* Boston, Mass.: Mariner, 2002.

Weber, David J. *The Spanish Frontier in North America.* New Haven, Conn.: Yale University Press, 1992.

Weisberger, Bernard A. *America Afire: Jefferson, Adams, and the First Contested Election*. New York: Morrow, 2000.

Wheelan, Joseph. *Jefferson's Vendetta: The Pursuit of Aaron Burr and the Judiciary.* New York: Carroll and Graf, 2005.

Wilentz, Sean. *The Rise of American Democracy: Jefferson to Lincoln*. New York: Norton, 2005.

Wills, Garry. *James Madison.* American Presidents Series. New York: Times Books, 2002.

Witham, Larry. *A City upon a Hill: How Sermons Changed the Course of American History.* New York: HarperCollins, 2007.

Zacks, Richard. *The Pirate Coast: Thomas Jefferson, the First Marines, and the Secret Mission of 1805.* New York: Hyperion, 2005.

Index

THE DON'T KNOW MUCH ABOUT® LIBRARY

FOR EVERYTHING YOU NEED TO KNOW BUT NEVER LEARNED

978-0-06-171980-6

978-0-06-008382-3

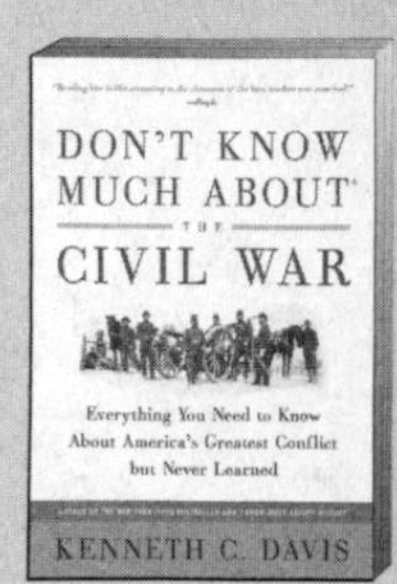

978-0-380-71908-2

978-0-380-72839-8

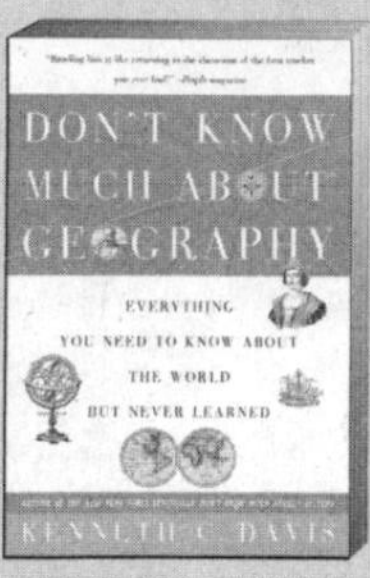

978-0-380-71379-0

978-0-06-093257-2

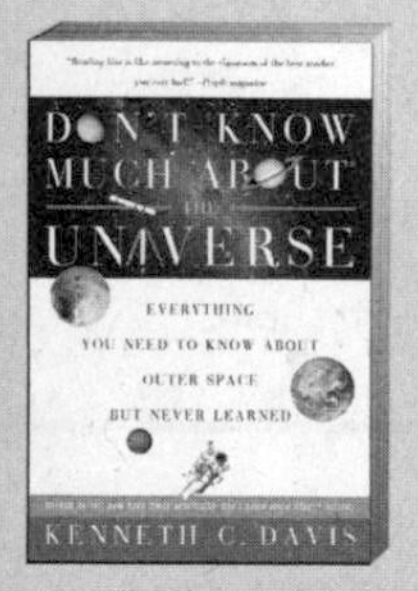

978-0-06-093256-5

978-0-06-125146-7

978-0-06-156232-7

ALSO BY KENNETH C. DAVIS

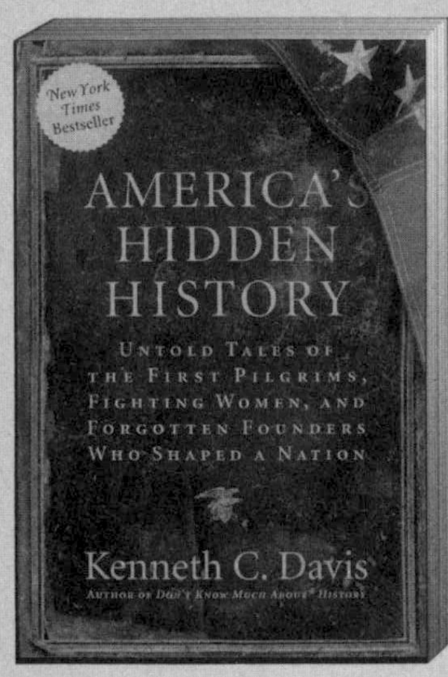

AMERICA'S HIDDEN HISTORY

UNTOLD TALES OF THE FIRST PILGRIMS, FIGHTING WOMEN, AND FORGOTTEN FOUNDERS WHO SHAPED A NATION

ISBN 978-0-06-111819-7 (paperback)

Kenneth C. Davis presents a collection of extraordinary stories, each detailing an overlooked episode that shaped the nation's destiny and character. Dramatic narratives set the record straight and bring to light fascinating facts from a time when the nation's fate hung in the balance.

"Once again Kenneth C. Davis proves that what you don't know can be shocking. Do yourself a favor. Read this book."
—Richard M. Cohen, author of *Strong at the Broken Places*

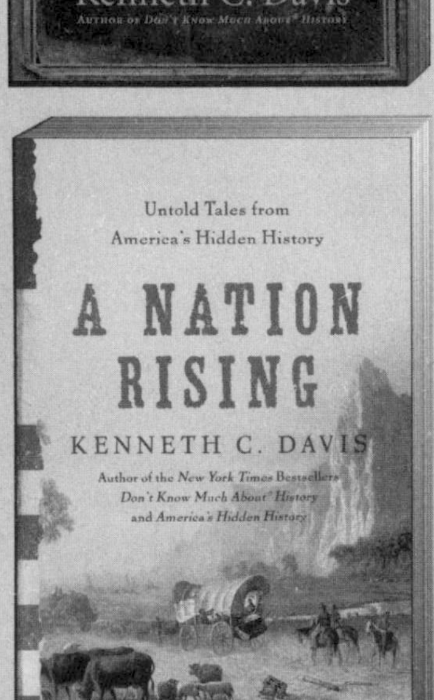

A NATION RISING

UNTOLD TALES FROM AMERICA'S HIDDEN HISTORY

ISBN 978-0-06-111821-0 (paperback)

Kenneth C. Davis explores the gritty first half of the 19th century, among the most tumultuous in the nation's short life, when the United States moved meteorically from a tiny, newborn nation, desperately struggling for survival on the Atlantic seaboard, to a near-empire that spanned the continent.

"Davis evokes the raw and violent spirit not just of an 'expanding nation,' but of an emerging and aggressive empire."
—Ray Raphael, author of *Founders*